A First Course on Orthogonal Polynomials

A First Course on Orthogonal Polynomials: Classical Orthogonal Polynomials and Related Topics provides an introduction to orthogonal polynomials and special functions aimed at graduate students studying these topics for the first time. A large part of its content is essentially inspired by the works of Pascal Maroni on the so-called algebraic theory of orthogonal polynomials, which distinguishes it from other contributions in the field.

Features

- Suitable for a graduate course in orthogonal polynomials
- Can be used for a short course on the algebraic theory of orthogonal polynomials and its applicability to the study of the "old" classical orthogonal polynomials
- Includes numerous exercises for each topic
- Real and complex analyses are the only prerequisites

A First Course on Orthogonal Polynomials

Classical Orthogonal Polynomials and Related Topics

Kenier Castillo and José Carlos Petronilho

CRC Press is an imprint of the
Taylor & Francis Group, an **informa** business
A CHAPMAN & HALL BOOK

Designed cover image: Catarina Sobral

First edition published 2025
by CRC Press
2385 NW Executive Center Drive, Suite 320, Boca Raton FL 33431

and by CRC Press
4 Park Square, Milton Park, Abingdon, Oxon, OX14 4RN

CRC Press is an imprint of Taylor & Francis Group, LLC

Library of Congress Cataloging-in-Publication Data
Names: Castillo, Kenier, author. | Petronilho, José Carlos, author.
Title: A first course on orthogonal polynomials : classical orthogonal polynomials and related topics / Kenier Castillo and José Carlos Petronilho.
Description: First edition. | Boca Raton : C&H/CRC Press, 2025. | Includes bibliographical references and index. | Summary: "A First Course on Orthogonal Polynomials: Classical Orthogonal Polynomials and Related Topics provides an introduction to orthogonal polynomials and special functions aimed at graduate students studying these topics for the first time. A large part of its content is essentially inspired by the works of Pascal Maroni on the so-called algebraic theory of orthogonal polynomials, which distinguishes it from other contributions in the field"-- Provided by publisher.
Identifiers: LCCN 2024022723 (print) | LCCN 2024022724 (ebook) | ISBN 9781032555973 (hardback) | ISBN 9781032560465 (paperback) | ISBN 9781003432067 (ebook)
Subjects: LCSH: Orthogonal polynomials.
Classification: LCC QA404.5 .C37 2025 (print) | LCC QA404.5 (ebook) | DDC 515/.55--dc23/eng/20240628
LC record available at https://lccn.loc.gov/2024022723
LC ebook record available at https://lccn.loc.gov/2024022724

ISBN: 978-1-032-55597-3 (hbk)
ISBN: 978-1-032-56046-5 (pbk)
ISBN: 978-1-003-43206-7 (ebk)

DOI: 10.1201/9781003432067

Typeset in Nimbus Roman
by KnowledgeWorks Global Ltd.

Publisher's note: This book has been prepared from camera-ready copy provided by the authors.

To the memory of the late Pascal Maroni,
in appreciation of his outstanding contributions

Contents

Preface

This book was written and designed for students who have never had previous contact with Special Functions and Orthogonal Polynomials. A graduate course on Real and Complex Analysis is a sufficient basis for the study of the material of this text. A large part of its content is essentially inspired by the works of Pascal Maroni on the so-called algebraic theory of orthogonal polynomials, which distinguishes it from other contributions in the field. The book can be shortened into a smaller course mainly centered on the algebraic theory of orthogonal polynomials and its applicability to the study of the "old" classical orthogonal polynomials—namely, Hermite, Laguerre, Jacobi, and Bessel polynomials—, by omitting the chapters 4, 5, 9, and 11, as well as the majority of Chapter 1. Chapter 9 contains a generalization of the results of Chapter 7, considering classical orthogonal polynomials on lattices, and it may require an additional comprehension effort from the students. It requires little effort to read Chapters 4, 5, and 11 independently.

The origin of the material included in this book began taking shape when I joined the Centre for Mathematics of the University of Coimbra (Portugal) in 2015. At that time, Petronilho and I started to work on some general questions about classical polynomials on lattices that can be found in Chapter 9 and which would later allow us to provide answers to several open problems in the area. It was during that time that we wrote a large survey article containing the results of Chapters 1–2, 6–8, and 10, which have recently appeared published in a shortened version. In the academic year 2015/2016, Petronilho taught a doctorate course on Special Functions and Orthogonal Polynomials in the PhD Program in Mathematics run by the Universities of Coimbra and Porto, and started writing class notes for his students which are at the core of this manuscript. These class notes, apart from containing, currently divided into chapters, everything that we had already written about classical orthogonal polynomials, were enriched with Chapters 3 (partially), 4, 5, and 11, which were written exclusively by Petronilho. In the academic years 2016/2017 and 2020/2021, Petronilho and I lectured the same course together, following his invitation to teach alongside him. We essentially used his class notes during the course, only adding a few exercises and correcting several typos thanks to the careful reading of students. A last revision by students was carried out in the academic year 2023/2024, as I used the final version of this book in a doctorate course on Special Functions and Orthogonal Polynomials that

I lectured in the PhD Program in Mathematics at the University of Cádiz (Spain).

After Petronilho sadly passed away, I sought to publish his class notes in book form. In order to bring this work up to the standard required for a book, I decided to carry out a deep revision of the text, adapt the content of several chapters, add a new chapter which included more recent results published together with Petronilho and assume full responsibility for all the content that we have written together, which consists of six chapters in its present version. The final result of my efforts was that I successfully submitted the new manuscript to CRC Press, Taylor & Francis Group so that it could sit alongside one of Petronilho's favorite mathematical books, "Measure Theory and Finite Properties of Functions" by Evans and Gariepy.

I would like to thank Pascal Maroni, who sadly passed away recently, for all his comments on the final version of this manuscript and for his extensive discussions about its contents during his last visit to Porto. I would also like to thank every student and colleague who read and commented on the content of this book. I must highlight three colleagues who deserve a special mention, Renato Álvarez Nodarse, José Luis Cardoso, and Márcio Nascimento de Jesus, who were the closest friends of Petronilho, for many useful comments. My ex-students Dieudonné Mbouna and Jorge Rivero Dones, and two current students, Guillermo Gordillo Nuñez and Alexandre Suzuki, read the final version of the manuscript, making suggestions for improvements. I also thank the three anonymous referees whose comments and remarks improved the initial version of the manuscript. I also want to thank Catarina Sobral for the lovely cover design. I am pleased to acknowledge the support of the Fundação para a Ciência e a Tecnologia (FCT) through DOI 10.54499/2022.00143.CEECIND/CP1714/CT0002 and the Centre for Mathematics of the University of Coimbra (funded by the Portuguese Government through FCT/MCTES, DOI 10.54499/UIDB/00324/2020).

I have decided, after obtaining the consent of Petronilho's wife, Célia, to renounce any material benefit derived from this publication and to relinquish all rights to Taylor & Francis Group to support *Research4Life*, which provides institutions in low-and middle-income countries with online access to academic and professional peer-reviewed content.

Kenier Castillo

Canidelo, Vila Nova de Gaia, May 2024

Authors

Kenier Castillo is currently the FCT Assistant Researcher at Centre for Mathematics of the University of Coimbra (Portugal). He earned his doctorate from Carlos III University of Madrid (Spain) in 2012. His research interests include Special Functions and Orthogonal Polynomials. He has more than 50 papers in peer-review journals. Castillo supervised two doctoral theses, one at the University of Coimbra and the other at the University of Cádiz (Spain), and currently supervises two PhD students at the University of Coimbra.

José Carlos Petronilho was an Associate Professor with Aggregation at the Department of Mathematics at the University of Coimbra. He earned his doctorate from the University of Coimbra in 1997. His research interests included Special Functions and Orthogonal Polynomials. He authored more than 40 papers in peer-review journals. In June 2024, a conference entitled "From Classical to Modern Analysis" within the framework of the European Congress of Mathematics was held to honor the memory of José Carlos Petronilho.

Symbol Description

Symbol	Description
$(a)_n$	$\Gamma(n+a)/\Gamma(a)$
$\{\mathbf{a}_n\}_{n\geq 0}$	dual basis
$B(\cdot,\cdot)$	Beta function
$\widehat{B}_n^{(\alpha)}$	monic Bessel polynomials
$C_n^{(a)}$	Charlier polynomials
C_n^{λ}	Gegenbauer (or ultraspherical) polynomials
$\mathbb{C}$	the complex numbers
D	derivative operator
D	x-derivative operator
$\mathbf{D}$	x-derivative in $\mathcal{P}'$
$D\mathbf{u}$	derivative of $\mathbf{u}$
$F(\cdot)$	Stieltjes's transform
g.c.d.	the greatest common divisor
h_a	homothetic operator
$\mathbf{h}_a$	dual operator of h_a
H_n	Hankel determinant
$H_n(\cdot)$	Hermite polynomials
$\widehat{H}_n$	monic Hermite polynomials
$\Im z$	the imaginary part of z
$L_n^{(\alpha)}$	Laguerre polynomials
$\widehat{L}_n^{(\alpha)}$	monic Laguerre polynomials
$m_n(\cdot;\beta,c)$	Meixner polynomials
$\mathbb{N}_0$	$\{0,1,\dots\}$
$\mathbb{N}$	$\{1,2,\dots\}$
$\mathcal{P}$	the set of all polynomials
$\mathcal{P}^*$	the algebraic dual of $\mathcal{P}$
$\mathcal{P}'$	the topological dual of $\mathcal{P}$
$\mathcal{P}_n$	the set of all polynomials with degree at most n
$\mathcal{P}_{-1}$	$\{0\}$
$\{P_n\}_{n\geq 0}$	orthogonal polynomial sequence (OPS) or Legendre polynomials
$\widehat{P}_n$	monic OPS
$P_n^{(\alpha,\beta)}$	Jacobi polynomials
$\widehat{P}_n^{(\alpha,\beta)}$	monic Jacobi polynomials
$\mathbb{Q}$	the rational numbers
$\Re z$	the real part of z
$\mathbb{R}$	the real numbers
$s_p(\mathbf{u})$	$\lvert\langle\mathbf{u},p\rangle\rvert$, $p\in\mathcal{P}$
sgn	signum function
S	x-average operator
$\mathbf{S}$	x-average in $\mathcal{P}'$
$S\mathbf{u}$	formal Stieltjes series associated with $\mathbf{u}$
$\mathrm{supp}(\mu)$	support of a measure μ
$\mathbb{S}^1$	$\{z\in\mathbb{C}:\lvert z\rvert=1\}$
$\mathbf{u}_B^{(\alpha)}$	Bessel functional
$\mathbf{u}_H$	Hermite functional
$\mathbf{u}_J^{(\alpha,\beta)}$	Jacobi functional
$\mathbf{u}_L^{(\alpha)}$	Laguerre functional
T_n	Chebyshev polynomials of second kind
U_n	Chebyshev polynomials of second kind
$\mathbf{u}\mathbf{v}$	product of $\mathbf{u}$ by $\mathbf{v}$
$\mathbf{u}\psi$	right multiplication of $\mathbf{u}$ by ψ
$\lvert\mathbf{u}\rvert_n$	$\max_{0\leq k\leq n}\lvert\langle\mathbf{u},x^k\rangle\rvert$

$\mathbb{Z}$	$\{\dots,-1,0,1,\dots\}$	τ_a	translator operator
$\mathbb{Z}^-$	$\{-1,-2,\dots\}$	$\boldsymbol{\tau}_a$	dual operator of τ_a
$(\alpha)_n$	Pochhammer symbol	$\phi\mathbf{u}$	left multiplication of $\mathbf{u}$ by ϕ
γ	Euler-Mascheroni constant	$\binom{n}{k}$	binomial coefficient
$\Gamma(z)$	Gamma function	$(x-c)^{-1}\mathbf{u}$	division of $\mathbf{u}$ by $x-c$
δ_c	the Dirac functional at a point c	$\#E$	cardinality of E
$\delta_{n,m}$	the Kronecker symbol	∇	the backward difference operator
Δ'	the set of all power series		
$\sigma(\psi)$	spectrum of a distribution function ψ		

Chapter 1

Foundations of the Algebraic Theory

In this chapter we give a brief introduction to the algebraic theory of orthogonal polynomials developed by Maroni, in connection with the ideas of Nikiforov, Suslov, and Uvarov on orthogonal polynomials on lattices or grids. We do not assume the reader has prior knowledge of locally convex spaces, so Appendix A contains some results that the reader should be familiar with before reading this chapter.

1.1 The Space $\mathcal{P}$

In his monumental work *"Une théorie algébrique des polynômes orthogonaux. Applications aux polynômes orthogonaux semiclassiques"* [34], Maroni presented in a definitive way the algebraic theory of Orthogonal Polynomials (OP) that he had begun to develop a few years earlier [31, 32, 33]. The essential idea is to consider OP as test functions living in an appropriate Locally Convex Space (LCS), say $\mathcal{P}$. This LCS is the set of all polynomials with real or complex coefficients endowed with a strict inductive limit topology:

$$\mathcal{P} = \bigcup_{n=0}^{\infty} \mathcal{P}_n = \operatorname{ind}\lim_n \mathcal{P}_n,$$

$\mathcal{P}_n$ being the space of all polynomials of degree at most n. For simplicity, in what follows, we do not distinguish between polynomial and polynomial function. In order to understand the theory that will be presented here, it is not essential to know much about LCS, but the reader should keep in mind that the reason why a topology is introduced is that it implies the following fundamental property:

Theorem 1.1. *Let $\mathcal{P} := \operatorname{ind}\lim_n \mathcal{P}_n$, and let $\mathcal{P}^*$ and $\mathcal{P}'$ be the algebraic and the topological duals of $\mathcal{P}$, respectively. Then*

$$\mathcal{P}' = \mathcal{P}^*. \tag{1.1}$$

DOI: 10.1201/9781003432067-1

Proof. Obviously, $\mathcal{P}' \subseteq \mathcal{P}^*$. By Theorem A.3, $\mathbf{u} \in \mathcal{P}'$ if the restriction $\mathbf{u}|\mathcal{P}_n$ is continuous for each n. But this is trivial because $\mathbf{u}|\mathcal{P}_n$ is a linear function defined on a finite-dimensional normed space. Thus $\mathcal{P}^* \subseteq \mathcal{P}'$, and the theorem follows. □

In what follows, we do not distinguish between $\mathcal{P}^*$ and $\mathcal{P}'$. The property (1.1) means that every linear function defined in $\mathcal{P}$ is continuous for the strict inductive limit topology in $\mathcal{P}$. This is a quite curious property, because given X is a normed vector space, $X' = X^*$ if $\dim X < \infty$, but $X' \neq X^*$ whenever $\dim X = \infty$. Of course, there is no contradiction between (1.1) and the fact that $\mathcal{P}$ is an infinite-dimensional vector space because $\mathcal{P}$ carried with the inductive limit topology is not a normed space. Indeed, $\mathcal{P}$ being a strict inductive limit of the spaces $\mathcal{P}_n$, and taking into account that each $\mathcal{P}_n$ is a proper closed subspace of $\mathcal{P}_{n+1}$—$\mathcal{P}$ is a hyperstrict inductive limit of the spaces $\mathcal{P}_n$—, the general theory of LCS (see Theorem A.4) ensures that $\mathcal{P}$ cannot be a metrizable space, and so *a fortiori* it is not a normed space. To be more precise, it is not possible to provide a norm that generates in $\mathcal{P}$ the above inductive limit topology. In $\mathcal{P}'$ we shall consider the weak dual topology, which, by definition, is generated by the family of seminorms $s_p : \mathcal{P}' \to [0, +\infty)$, defined by

$$s_p(\mathbf{u}) := |\langle \mathbf{u}, p \rangle|, \quad p \in \mathcal{P}. \tag{1.2}$$

(Recall that for $p \in \mathcal{P}$ and $\mathbf{u} \in \mathcal{P}^*$, the element $\mathbf{u}(p)$ is denoted by $\langle \mathbf{u}, p \rangle$.) It turns out that this family of seminorms s_p is equivalent to the family of seminorms $|\cdot|_n : \mathcal{P}' \to [0, +\infty)$, defined by

$$|\mathbf{u}|_n := \max_{0 \leq k \leq n} |\langle \mathbf{u}, x^k \rangle|, \quad n \in \mathbb{N}_0. \tag{1.3}$$

The relation between s_p and $|\cdot|_n$ is established by the next theorem.

Theorem 1.2. *Define* $\mathcal{S} := \{s_p : p \in \mathcal{P}\}$ *and* $\mathcal{S}_\sharp := \{|\cdot|_n : n \in \mathbb{N}_0\}$, *with* s_p *and* $|\cdot|_n$ *given by* (1.2) *and* (1.3), *respectively. Then* $\mathcal{S}$ *and* $\mathcal{S}_\sharp$ *are equivalent families of seminorms in* $\mathcal{P}'$.

Proof. Set $p(x) := \sum_{j=0}^n a_j x^j$ and $C(p) := \sum_{j=0}^n |a_j|$, $n \in \mathbb{N}_0$. Note that

$$s_p(\mathbf{u}) = |\langle \mathbf{u}, p \rangle| = \left| \sum_{j=0}^n a_j \langle \mathbf{u}, x^j \rangle \right| \leq C(p) |\mathbf{u}|_n.$$

On the other hand, setting $p_j(x) := x^j$, $j \in \mathbb{N}_0$, we have

$$|\mathbf{u}|_n = \max_{0 \leq j \leq n} |\langle \mathbf{u}, x^j \rangle| \leq \sum_{j=0}^n |\langle \mathbf{u}, x^j \rangle| = \sum_{j=0}^n s_{p_j}(\mathbf{u}),$$

and, by Proposition A.2, the result follows. □

Since $\mathcal{S}_\sharp$ is a countable family of seminorms, by Theorem 1.2 and Theorem A.1, $\mathcal{P}'$ is a metrizable space with a metric

$$\varrho(\mathbf{u},\mathbf{v}) := \sum_{n=0}^{\infty} \frac{1}{2^n} \frac{|\mathbf{u}-\mathbf{v}|_n}{1+|\mathbf{u}-\mathbf{v}|_n}, \quad \mathbf{u},\mathbf{v} \in \mathcal{P}'. \tag{1.4}$$

In addition, $\mathcal{P}'$ is a Fréchet space.

1.2 Dual Basis

Since we will work in the dual space $\mathcal{P}'$, it would be useful to explicitly build bases therein. This makes sense since (1.1) allows us to write expansions (finite or infinite sums) of the elements of $\mathcal{P}'$ in terms of the elements of a given basis, in the sense of the weak dual topology in $\mathcal{P}'$. Such basis in $\mathcal{P}'$ may be achieved in a natural way, using simple sets of polynomials. A simple set in $\mathcal{P}$ is a sequence of nonzero polynomials, say $\{R_n\}_{n\geq 0}$, such that $\deg R_n = n$. To any simple set $\{R_n\}_{n\geq 0}$, we may associate a dual basis which by definition is a sequence of linear functionals $\{\mathbf{a}_n\}_{n\geq 0}$, $\mathbf{a}_n : \mathcal{P} \to \mathbb{C}$, such that

$$\langle \mathbf{a}_n, R_k \rangle := \delta_{n,k}, \quad k \in \mathbb{N}_0, \tag{1.5}$$

$\delta_{n,k}$ being the Kronecker symbol: $\delta_{n,k} = 1$ if $n = k$; $\delta_{n,k} = 0$ if $n \neq k$. The following theorem plays a fundamental role in the sequel.

Theorem 1.3. *Let $\{R_n\}_{n\geq 0}$ be a simple set in $\mathcal{P}$, and let $\{\mathbf{a}_n\}_{n\geq 0}$ be the associated dual basis. Then*

$$\mathbf{u} = \sum_{n=0}^{\infty} \langle \mathbf{u}, R_n \rangle \, \mathbf{a}_n, \quad \mathbf{u} \in \mathcal{P}', \tag{1.6}$$

in the sense of the weak dual topology in $\mathcal{P}'$.

Proof. Fix $N \in \mathbb{N}$ and let

$$\mathbf{s}_N := \sum_{n=0}^{N-1} \langle \mathbf{u}, R_n \rangle \mathbf{a}_n,$$

be the partial sum of order N of the series appearing in (1.6). We need to show that

$$\lim_{N\to\infty} \langle \mathbf{s}_N - \mathbf{u}, p \rangle = 0, \quad p \in \mathcal{P}.$$

Clearly, it suffices to prove that this equality holds for $p \in \{R_0, R_1, R_2, \dots\}$. Indeed, for $N > k$ and $k \in \mathbb{N}_0$ fixed,

$$\langle \mathbf{s}_N - \mathbf{u}, R_k \rangle = \sum_{n=0}^{N-1} \langle \mathbf{u}, R_n \rangle \langle \mathbf{a}_n, R_k \rangle - \langle \mathbf{u}, R_k \rangle = 0,$$

and so

$$\lim_{N\to\infty} \langle \mathbf{s}_N - \mathbf{u}, R_k \rangle = 0,$$

which proves the theorem. □

1.3 Basic Operations

Given $\mathbf{u} \in \mathcal{P}'$, we will denote by

$$u_n := \langle \mathbf{u}, x^n \rangle, \quad n \in \mathbb{N}_0,$$

the moment of order n of $\mathbf{u}$. Clearly, if $\mathbf{u}$ and $\mathbf{v}$ are two elements in $\mathcal{P}'$ such that the corresponding moments satisfy $u_n = v_n$ for each $n \in \mathbb{N}_0$, then $\mathbf{u} = \mathbf{v}$. In other words, each functional $\mathbf{u} \in \mathcal{P}'$ is uniquely determined by its sequence of moments. Define the operators M_ϕ, T, and θ_c, from $\mathcal{P}$ into $\mathcal{P}$, by

$$M_\phi p(x) := \phi(x)p(x), \quad Tp(x) := -p'(x), \tag{1.7}$$

$$\theta_c p(x) := \frac{p(x) - p(c)}{x - c}, \quad \phi, p \in \mathcal{P},\ c \in \mathbb{C}, \tag{1.8}$$

where $'$ denotes the standard derivative with respect to x. Note that $\theta_c p(x)$ is defined as above if $x \neq c$, with the obvious definition

$$\theta_c p(c) := p'(c) \quad \text{if} \quad x = c.$$

By Theorem A.5, the dual operators M'_ϕ, T', and θ'_c belong to $\mathcal{L}(\mathcal{P}', \mathcal{P}')$. For each $\mathbf{u} \in \mathcal{P}'$, the images $M'_\phi\mathbf{u}$, $T'\mathbf{u}$, and $\theta'_c\mathbf{u}$, are elements in $\mathcal{P}'$, hereafter denoted by $\phi\,\mathbf{u}$, $D\mathbf{u}$, and $(x - c)^{-1}\mathbf{u}$.

Definition 1.1. *Let $\mathbf{u} \in \mathcal{P}'$, $\phi, p \in \mathcal{P}$, and $c \in \mathbb{C}$.*

(i) *The left multiplication of $\mathbf{u}$ by ϕ, denoted by $\phi\mathbf{u}$, is the functional in $\mathcal{P}'$ defined by*

$$\langle \phi\mathbf{u}, p \rangle := \langle \mathbf{u}, \phi p \rangle;$$

(ii) *the derivative of* $\mathbf{u}$, *denoted by* $D\mathbf{u}$, *is the functional in* $\mathcal{P}'$ *defined by*

$$\langle D\mathbf{u}, p\rangle := -\langle \mathbf{u}, p'\rangle;$$

(iii) *the division of* $\mathbf{u}$ *by* $x - c$, *denoted by* $(x-c)^{-1}\mathbf{u}$, *is the functional in* $\mathcal{P}'$ *defined by*

$$\langle (x-c)^{-1}\mathbf{u}, p\rangle := \langle \mathbf{u}, \theta_c p\rangle = \left\langle \mathbf{u}, \frac{p(x)-p(c)}{x-c}\right\rangle.$$

The above definitions, introduced by duality with respect to the operators defined in (1.7) and (1.8), are in accordance with those usually given in Theory of Distributions.

Proposition 1.1. *Let* $\mathbf{u} \in \mathcal{P}'$ *and* $\phi \in \mathcal{P}$. *Then*

$$D(\phi\mathbf{u}) = \phi'\mathbf{u} + \phi\, D\mathbf{u}.$$

Proof. For each $p \in \mathcal{P}$, we get

$$\begin{aligned}\langle D(\phi\mathbf{u}), p\rangle &= -\langle \phi\mathbf{u}, p'\rangle = -\langle \mathbf{u}, \phi p'\rangle = -\langle \mathbf{u}, -\phi' p + (\phi p)'\rangle \\ &= \langle \mathbf{u}, \phi' p\rangle - \langle \mathbf{u}, (\phi p)'\rangle = \langle \phi'\mathbf{u}, p\rangle + \langle D\mathbf{u}, \phi p\rangle \\ &= \langle \phi'\mathbf{u} + \phi\, D\mathbf{u}, p\rangle,\end{aligned}$$

and the desired equality follows. □

Let us now introduce a generalization of the derivative operator D [42, 43, 44, 45], including, for instance, the Askey-Wilson operator. Fix $q > 0$. From now on, a lattice $x : \mathbb{C} \to \mathbb{C}$ is given by

$$x(s) := \begin{cases} \mathfrak{c}_1 q^{-s} + \mathfrak{c}_2 q^{s} + \mathfrak{c}_3, & q \neq 1, \\ \mathfrak{c}_4 s^2 + \mathfrak{c}_5 s + \mathfrak{c}_6, & q = 1, \end{cases} \tag{1.9}$$

where $\mathfrak{c}_1, \ldots, \mathfrak{c}_6$ are complex numbers, that may depend on q, such that $(\mathfrak{c}_1, \mathfrak{c}_2) \neq (0,0)$ if $q \neq 1$ and $(\mathfrak{c}_4, \mathfrak{c}_5, \mathfrak{c}_6) \neq (0,0,0)$ if $q = 1$. In the case $q = 1$, the lattice is called quadratic if $\mathfrak{c}_4 \neq 0$ and linear if $\mathfrak{c}_4 = 0$. If $q \neq 1$, the lattice is called q-quadratic if $\mathfrak{c}_1\mathfrak{c}_2 \neq 0$, and q-linear if $\mathfrak{c}_1\mathfrak{c}_2 = 0$. Note that

$$\frac{x(s+1/2) + x(s-1/2)}{2} = \alpha\, x(s) + \beta,$$

where α and β are given by

$$\alpha := \frac{q^{1/2}+q^{-1/2}}{2}, \qquad \beta := \begin{cases} (1-\alpha)\mathfrak{c}_3, & q \neq 1, \\ \mathfrak{c}_4/4, & q = 1. \end{cases} \tag{1.10}$$

The next fundamental theorem was proved in [4].

Theorem 1.4. *Let $x(s)$ be a lattice given by (1.9). Then*

$$\frac{x(s+n)+x(s)}{2} = \alpha_n x_n(s) + \beta_n,$$

$$x(s+n) - x(s) = \gamma_n \nabla x_{n+1}(s),$$

where

$$x_\mu(s) := x\left(s+\frac{\mu}{2}\right), \qquad \nabla f(s) := f(s) - f(s-1),$$

and $(\alpha_n)_{n\geq 0}$, $(\beta_n)_{n\geq 0}$, and $(\gamma_n)_{n\geq 0}$ are sequences of numbers generated by the following system of difference equations

$$\begin{array}{lll} \alpha_0 = 1, & \alpha_1 = \alpha, & \alpha_{n+1} - 2\alpha\alpha_n + \alpha_{n-1}6 = 0, \\ \beta_0 = 0, & \beta_1 = \beta, & \beta_{n+1} - 2\beta_n + \beta_{n-1} = 2\beta\alpha_n, \\ \gamma_0 = 0, & \gamma_1 = 1, & \gamma_{n+1} - \gamma_{n-1} = 2\alpha_n. \end{array}$$

Moreover, the explicit solutions of these difference equations are

$$\alpha_n = \frac{q^{n/2}+q^{-n/2}}{2}, \qquad \beta_n = \begin{cases} \beta\left(\dfrac{q^{n/4}-q^{-n/4}}{q^{1/4}-q^{-1/4}}\right)^2, & q \neq 1, \\ \beta n^2, & q = 1, \end{cases}$$

$$\gamma_n = \begin{cases} \dfrac{q^{n/2}-q^{-n/2}}{q^{1/2}-q^{-1/2}}, & q \neq 1, \\ n, & q = 1. \end{cases}$$

We also point out the following useful relations:

Proposition 1.2. *Set $\alpha_{-1} = \alpha$ and $\gamma_{-1} = \gamma$. Under the assumptions of*

Theorem 1.4, the following properties hold:

$$\gamma_{n+1} - 2\alpha\gamma_n + \gamma_{n-1} = 0, \tag{1.11}$$

$$\alpha_n + \gamma_{n-1} = \alpha\gamma_n, \tag{1.12}$$

$$(2\alpha^2 - 1)\alpha_n + (\alpha^2 - 1)\gamma_{n-1} = \alpha\alpha_{n+1}, \tag{1.13}$$

$$\gamma_{2n} = 2\alpha_n\gamma_n, \tag{1.14}$$

$$\alpha_n^2 + (\alpha^2 - 1)\gamma_n^2 = \alpha_{2n} = 2\alpha_n^2 - 1, \tag{1.15}$$

$$\alpha_{n-1} - \alpha\,\alpha_n = (1 - \alpha^2)\gamma_n, \tag{1.16}$$

$$\alpha + \alpha_n\gamma_n = \alpha_{n-1}\gamma_{n+1}, \tag{1.17}$$

$$1 + \alpha_{n+1}\gamma_n = \alpha_n\gamma_{n+1}. \tag{1.18}$$

The following definition appears in [43, (3.2.4)–(3.2.5)] up to a shift in the variable s.

Definition 1.2. *Let $x(s)$ be a lattice given by (1.9) and let $p \in \mathcal{P}$. The x-derivative and the x-average (or simply derivative and average) denoted, respectively, by* D *and* S*, are defined by* $\deg(\mathrm{D}p) := \deg p - 1$, $\deg(\mathrm{S}p) := \deg p$*, and*

$$\mathrm{D}p(x(s)) := \frac{p(x(s+1/2)) - p(x(s-1/2))}{x(s+1/2) - x(s-1/2)},$$

$$\mathrm{S}p(x(s)) := \frac{p(x(s+1/2)) + p(x(s-1/2))}{2},$$

where $\mathrm{D}p := p'$ *and* $\mathrm{S}p := p$ *whenever* $x(s) = \mathfrak{c}_6$.

As for the case of the standard derivation, the operators D and S induce two operations on $\mathcal{P}'$ [17].

Definition 1.3. *Let $x(s)$ be a lattice given by (1.9). Let $p \in \mathcal{P}$ and let $\mathbf{u} \in \mathcal{P}'$. The x-derivative and the x-average of $\mathbf{u}$ (or simply derivative and average of $\mathbf{u}$, if no confusion seems likely to arise) denoted, respectively,* *by* **Du** *and* **Su**, *are the functionals in* $\mathcal{P}'$ *defined by*

$$\langle \mathbf{Du}, p \rangle := -\langle \mathbf{u}, \mathrm{D}p \rangle, \quad \langle \mathbf{Su}, p \rangle := \langle \mathbf{u}, \mathrm{S}p \rangle.$$

We consider two fundamental polynomials, $\mathtt{U}_1$ and $\mathtt{U}_2$, defined by

$$\mathtt{U}_1(z) := (\alpha^2 - 1)z + \beta(\alpha + 1), \tag{1.19}$$

$$\mathtt{U}_2(z) := (\alpha^2 - 1)z^2 + 2\beta(\alpha + 1)z + \delta, \tag{1.20}$$

where $\delta \equiv \delta_x$ is a constant with respect to the lattice, given by

$$\delta := \left(\frac{x(0) + x(1) - 2\beta(\alpha + 1)}{2\alpha} \right)^2 - x(0)x(1). \tag{1.21}$$

The following expressions play a fundamental role in this context:

$$\delta = \begin{cases} (\alpha^2 - 1)(\mathfrak{c}_3^2 - 4\mathfrak{c}_1\mathfrak{c}_2), & q \neq 1, \\ \mathfrak{c}_5^2/4 - \mathfrak{c}_4\mathfrak{c}_6, & q = 1, \end{cases}$$

$$\mathtt{U}_1(x(s)) = \begin{cases} (\alpha^2 - 1)(x(s) - \mathfrak{c}_3), & q \neq 1, \\ \mathfrak{c}_4/2, & q = 1, \end{cases}$$

$$\mathtt{U}_2(x(s)) = \begin{cases} (\alpha^2 - 1)((x(s) - \mathfrak{c}_3)^2 - 4\mathfrak{c}_1\mathfrak{c}_2), & q \neq 1, \\ \mathfrak{c}_4(x(s) - \mathfrak{c}_6) + \dfrac{1}{4}\mathfrak{c}_5^2, & q = 1. \end{cases}$$

Finally, we recall the following useful relations which can be easily proved:

$$\mathrm{D}\mathtt{U}_1 = \alpha^2 - 1, \qquad \mathrm{S}\mathtt{U}_1 = \alpha\mathtt{U}_1, \tag{1.22}$$

$$\mathrm{D}\mathtt{U}_2 = 2\alpha\mathtt{U}_1, \qquad \mathrm{S}\mathtt{U}_2 = \alpha^2\mathtt{U}_2 + \mathtt{U}_1^2. \tag{1.23}$$

The next identities are proved in [40] and [10]. We prove several of them here and leave the rest to the readers.

Proposition 1.3. *Let* $\mathbf{u} \in \mathcal{P}'$ *and* $p, r \in \mathcal{P}$. *Then the following properties* *hold:*

$$\mathrm{D}(pr) = (\mathrm{D}p)(\mathrm{S}r) + (\mathrm{S}p)(\mathrm{D}r), \tag{1.24}$$

$$\mathrm{S}^2 p = \alpha^{-1}\mathrm{S}(\mathcal{U}_1 \mathrm{D}p) + \alpha^{-1}\mathcal{U}_2 \mathrm{D}^2 p + p, \tag{1.25}$$

$$\mathbf{D}(p\mathbf{u}) = \big(\mathrm{S}p - \alpha^{-1}\mathcal{U}_1 \mathrm{D}p\big)\,\mathbf{Du} + \alpha^{-1}(\mathrm{D}p)\mathbf{Su}, \tag{1.26}$$

$$\mathbf{S}(p\mathbf{u}) = \big(\mathrm{S}p - \alpha^{-1}\mathcal{U}_1 \mathrm{D}p\big)\,\mathbf{Su} + \alpha^{-1}(\mathrm{D}p)\mathbf{D}(\mathcal{U}_2\mathbf{u}), \tag{1.27}$$

$$\mathbf{S}(p\mathbf{u}) = (\alpha\mathcal{U}_2 - \alpha^{-1}\mathcal{U}_1^2)(\mathrm{D}p)\mathbf{Du} + (\mathrm{S}p + \alpha^{-1}\mathcal{U}_1 \mathrm{D}p)\mathbf{Su}, \tag{1.28}$$

$$\mathbf{D}^2(\mathcal{U}_2\mathbf{u}) = \alpha\mathbf{S}^2\mathbf{u} + \mathbf{D}(\mathcal{U}_1\mathbf{Su}) - \alpha\mathbf{u}, \tag{1.29}$$

$$\mathbf{D}^2(\mathcal{U}_2\mathbf{u}) = (2\alpha - \alpha^{-1})\mathbf{S}^2\mathbf{u} + \alpha^{-1}\mathcal{U}_1\mathbf{DSu} - \alpha\mathbf{u}. \tag{1.30}$$

Proof. To prove (1.28), set $p = \mathtt{U}_2$ in (1.26) and then use $\mathrm{S}\mathtt{U}_2 = \alpha^2\mathtt{U}_2 + \mathtt{U}_1^2$ and $\mathrm{D}\mathtt{U}_2 = 2\alpha\mathtt{U}_1$ to obtain

$$\begin{aligned}\mathbf{D}(\mathtt{U}_2\mathbf{u}) &= \big(\mathrm{S}\mathtt{U}_2 - \alpha^{-1}\mathtt{U}_1\mathrm{D}\mathtt{U}_2\big)\,\mathbf{Du} + \alpha^{-1}(\mathrm{D}\mathtt{U}_2)\mathbf{Su} \\ &= (\alpha^2\mathtt{U}_2 - \mathtt{U}_1^2)\mathbf{Du} + 2\mathtt{U}_1\mathbf{Su}.\end{aligned}$$

Replacing this expression in the right-hand side of (1.27) we obtain (1.28). Next, taking arbitrarily $p \in \mathcal{P}$, we have

$$\begin{aligned}\langle \mathbf{D}^2(\mathtt{U}_2\mathbf{u}), p\rangle &= \langle \mathbf{u}, \mathtt{U}_2\mathrm{D}^2 p\rangle = \langle \mathbf{u}, \alpha\mathrm{S}^2 p - \mathrm{S}(\mathtt{U}_1\mathrm{D}p) - \alpha p\rangle \\ &= \langle \alpha\mathbf{S}^2\mathbf{u} + \mathbf{D}(\mathtt{U}_1\mathbf{Su}) - \alpha\mathbf{u}, p\rangle,\end{aligned}$$

where the second equality holds by (1.25). This proves (1.29). Setting $p = \mathtt{U}_1$ in (1.26) and replacing therein $\mathbf{u}$ by $\mathbf{Su}$, and taking into account $\mathrm{S}\mathtt{U}_1 = \alpha\mathtt{U}_1$, we deduce

$$\mathbf{D}\big(\mathtt{U}_1\mathbf{Su}\big) = \alpha^{-1}\mathtt{U}_1\mathbf{DSu} + (\alpha - \alpha^{-1})\mathbf{S}^2\mathbf{u}.$$

Substituting this into the right-hand side of (1.29) we obtain (1.30). □

Remark 1.1. *Note that Proposition 1.1 is a particular case of* (1.26). *Indeed, taking* $x(s) = \mathfrak{c}_6$ ($q = 1$ *and* $\mathfrak{c}_4 = \mathfrak{c}_5 = 0$), *we have* $\mathrm{S}p = p$, $\alpha = 1$, $\mathcal{U}_1 = 0$, $\mathrm{D}p = p'$, *and* $\mathbf{Su} = \mathbf{u}$.

By definition, the left multiplication of a functional in $\mathcal{P}'$ by a polynomial is an element in $\mathcal{P}'$. We may also define the right multiplication of a linear function by a polynomial, but the result in this case will be a polynomial.

Definition 1.4. *Let* $\mathbf{u} \in \mathcal{P}'$ *and* $\psi \in \mathcal{P}$. *The right multiplication of* $\mathbf{u}$ *by* ψ, *denoted by* $\mathbf{u}\psi$, *is the polynomial defined by*

$$\mathbf{u}\psi(x) := \langle \mathbf{u}_\xi, \theta_\xi(x\psi) \rangle = \left\langle \mathbf{u}_\xi, \frac{x\psi(x) - \xi\psi(\xi)}{x - \xi} \right\rangle,$$

where the subscript ξ *in* $\mathbf{u}_\xi$ *means that this functional acts on polynomials in* ξ.

Setting $\psi(x) := \sum_{i=0}^{n} a_i x^i$, the polynomial $\mathbf{u}\psi$ is explicitly given by

$$\mathbf{u}\psi(x) = \sum_{i=0}^{n} \left(\sum_{j=i}^{n} a_j u_{j-i} \right) x^i, \tag{1.31}$$

and it also admits the following useful matrix representation:

$$\mathbf{u}\psi(x) = [a_0 \; a_1 \cdots a_n] \begin{bmatrix} u_0 & 0 & 0 & \cdots & 0 \\ u_1 & u_0 & 0 & \cdots & 0 \\ u_2 & u_1 & u_0 & \cdots & 0 \\ \vdots & \vdots & \vdots & \ddots & \vdots \\ u_n & u_{n-1} & u_{n-2} & \cdots & u_0 \end{bmatrix} \begin{bmatrix} 1 \\ x \\ x^2 \\ \vdots \\ x^n \end{bmatrix}. \tag{1.32}$$

The right multiplication of a functional by a polynomial enables us to introduce by duality a product in $\mathcal{P}'$. Indeed, fix $\mathbf{u} \in \mathcal{P}'$, and let $T_\mathbf{u} : \mathcal{P} \to \mathcal{P}$ be defined by

$$T_\mathbf{u}\, p := \mathbf{u}\, p.$$

The dual operator, $T'_\mathbf{u} : \mathcal{P}' \to \mathcal{P}'$, is given by

$$\langle T'_\mathbf{u}\mathbf{v}, p \rangle := \langle \mathbf{v}, T_\mathbf{u} p \rangle = \langle \mathbf{v}, \mathbf{u}p \rangle, \quad p \in \mathcal{P}.$$

Thus, we may introduce a product in $\mathcal{P}'$, by duality with respect to the right multiplication of a functional by a polynomial.

Definition 1.5. *Let* $\mathbf{u}, \mathbf{v} \in \mathcal{P}'$ *and* $p \in \mathcal{P}$. *The product* $\mathbf{uv}$ *is the functional in* $\mathcal{P}'$ *given by*

$$\langle \mathbf{uv}, p \rangle := \langle \mathbf{v}, \mathbf{u}p \rangle.$$

Note that $\mathbf{uv} = \mathbf{vu}$. This fact may be easily seen by noticing that the moments of the functionals $\mathbf{uv}$ and $\mathbf{vu}$ coincide. In fact,

$$\langle \mathbf{uv}, x^n \rangle = \sum_{i+j=n} u_i v_j = \langle \mathbf{vu}, x^n \rangle, \quad n \in \mathbb{N}_0. \tag{1.33}$$

Further, there exists a unit element in $\mathcal{P}'$, namely the Dirac functional at the origin δ_0. Indeed, using (1.33), it is easy to prove that

$$\mathbf{u}\delta_0 = \mathbf{u}, \quad \mathbf{u} \in \mathcal{P}'.$$

Recall that the Dirac functional at a point $c \in \mathbb{C}$, $\delta_c : \mathcal{P} \to \mathbb{C}$, is defined by

$$\langle \delta_c, p \rangle := p(c), \quad p \in \mathcal{P}.$$

Note that the vector space $\mathcal{P}'$ possessing the Cauchy product and unit element δ_0 is an algebra. The next theorem lists some basic properties concerning the above operations. The proof is left to the reader.

Proposition 1.4. *Let* $\mathbf{u}, \mathbf{v}, \mathbf{w} \in \mathcal{P}'$, $p \in \mathcal{P}$, *and* $c \in \mathbb{C}$. *Then the following properties hold:*

$$\delta_0 p = p, \qquad p(\mathbf{u}\mathbf{v}) = (p\mathbf{u})\mathbf{v} + x(\mathbf{u}\theta_0 p)\mathbf{v},$$

$$\mathbf{v}(\mathbf{u}p) = (\mathbf{v}\mathbf{u})p, \qquad \mathbf{u} = (x-c)\big((x-c)^{-1}\mathbf{u}\big),$$

$$(\mathbf{u}+\mathbf{v})\mathbf{w} = \mathbf{u}\mathbf{w} + \mathbf{v}\mathbf{w}, \qquad \mathbf{u} - u_0\delta_c = (x-c)^{-1}\big((x-c)\mathbf{u}\big),$$

$$(\mathbf{u}\mathbf{v})\mathbf{w} = \mathbf{u}(\mathbf{v}\mathbf{w}),$$

$\mathbf{u}$ *has an inverse iff* $u_0 \neq 0$.

On the course of the proof of several properties listed in Proposition 1.4, it may be useful to use the following identities, valid for any array $\{\alpha_{i,j}\}_{0\leq j\leq i\leq n}$ of $(n+1)(n+2)/2$ complex numbers:

$$\sum_{i=0}^{n}\sum_{j=0}^{i}\alpha_{i,j} = \sum_{i=0}^{n}\sum_{j=i}^{n}\alpha_{j,j-i} = \sum_{j=0}^{n}\sum_{i=j}^{n}\alpha_{i,j}.$$

Indeed, arrange the array elements so that they form a right triangle:

$$\begin{array}{lllll}
\alpha_{0,0} & & & & \\
\alpha_{1,0} & \alpha_{1,1} & & & \\
\alpha_{2,0} & \alpha_{2,1} & \alpha_{2,2} & & \\
\vdots & \vdots & \vdots & \ddots & \\
\alpha_{n,0} & \alpha_{n,1} & \alpha_{n,2} & \cdots & \alpha_{n,n}
\end{array}$$

Then we only need to notice that the first double sum is obtained by adding the elements by horizontal lines, from top to bottom, the second sum is obtained by adding the elements by diagonal lines, starting from the "hypotenuse", and the third one is obtained by adding the elements by vertical lines, from left to right. Note that from the last property listed in Proposition 1.4, we have

$$(x-c)\mathbf{u} = (x-c)\mathbf{v} \quad \text{iff} \quad \mathbf{u} = \mathbf{v} + (u_0 - v_0)\delta_c.$$

We conclude this section with the following theorem.

Theorem 1.5. *Let $\mathbf{u} \in \mathcal{P}'$ and $p, q \in \mathcal{P} \setminus \{0\}$, and denote by Z_p and Z_q the zeros of p and q, respectively. Then the following property holds:*

$$Z_p \cap Z_q = \emptyset \quad \wedge \quad p\mathbf{u} = q\mathbf{u} = \mathbf{0} \quad \Rightarrow \quad \mathbf{u} = \mathbf{0}.$$

Proof. Assume that $p\mathbf{u} = q\mathbf{u} = \mathbf{0}$. Since p and q are coprime, there exist polynomials a, b for which $a\,p + b\,q = 1$. Since $\langle p\mathbf{u}, a\,x^n\rangle = \langle q\mathbf{u}, b\,x^n\rangle = 0$ for all $n \in \mathbb{N}_0$, we see that

$$\langle \mathbf{u}, x^n\rangle = \langle \mathbf{u}, (a\,p + b\,q)x^n\rangle = 0,$$

and the result follows. □

1.4 The Leibniz Formula

In this section, we state a useful version of the Leibniz formula, but first, we need to state some preliminary results.

Lemma 1.1. *Let $p \in \mathcal{P}$. Then*

$$\mathrm{D}^n\mathrm{S}p = \alpha_n \mathrm{S}\mathrm{D}^n p + \gamma_n \mathrm{U}_1 \mathrm{D}^{n+1} p, \quad n \in \mathbb{N}_0. \tag{1.34}$$

Proof. The proof is by induction on n. Clearly, (1.34) holds for $n = 0$. Suppose that (1.34) holds for all positive integers less than or equal to a fixed n. By using (1.34) for n and $n = 1$ with p replaced by $\mathrm{D}^n p$, and applying (1.24) to

$\mathrm{D}(\mathtt{U}_1\mathrm{D}^{n+1}p)$ we deduce

$$\begin{aligned}\mathrm{D}^{n+1}\mathrm{S}p &= \mathrm{D}\left(\mathrm{D}^n\mathrm{S}p\right)\\ &= \mathrm{D}\left(\alpha_n\mathrm{S}\mathrm{D}^n p + \gamma_n\mathtt{U}_1\mathrm{D}^{n+1}p\right) = \alpha_n\mathrm{D}\mathrm{S}(\mathrm{D}^n p) + \gamma_n\mathrm{D}\left(\mathtt{U}_1\mathrm{D}^{n+1}p\right)\\ &= \alpha_n\left(\alpha\mathrm{S}\mathrm{D}^{n+1}p + \mathtt{U}_1\mathrm{D}^{n+2}p\right) + \gamma_n\left((\alpha^2-1)\mathrm{S}\mathrm{D}^{n+1}p + \alpha\mathtt{U}_1\mathrm{D}^{n+2}p\right)\\ &= \left(\alpha\alpha_n + (\alpha^2-1)\gamma_n\right)\mathrm{S}\mathrm{D}^{n+1}p + (\alpha_n + \alpha\gamma_n)\mathtt{U}_1\mathrm{D}^{n+2}p.\end{aligned}$$

Finally, using Proposition 1.2 we see that (1.34) holds whenever n is replaced by $n+1$, and the result follows. □

The next result is a functional version of the above lemma. The details are left to the reader.

Lemma 1.2. *Let* $\mathbf{u} \in \mathcal{P}'$. *Then*

$$\alpha\mathbf{D}^n\mathbf{S}\mathbf{u} = \alpha_{n+1}\mathbf{S}\mathbf{D}^n\mathbf{u} + \gamma_n\mathtt{U}_1\mathbf{D}^{n+1}\mathbf{u}, \quad n \in \mathbb{N}_0. \tag{1.35}$$

Let us introduce an operator $\mathrm{T}_{n,k} : \mathcal{P} \to \mathcal{P}$, $n, k \in \mathbb{N}_0$, defined as follows: If $n = k = 0$, we set

$$\mathrm{T}_{0,0}p := p; \tag{1.36}$$

and if $n \geq 1$ and $0 \leq k \leq n$, we set

$$\mathrm{T}_{n,k}p := \mathrm{S}\mathrm{T}_{n-1,k}p - \frac{\gamma_{n-k}}{\alpha_{n-k}}\mathtt{U}_1\mathrm{D}\mathrm{T}_{n-1,k}p + \frac{1}{\alpha_{n+1-k}}\mathrm{D}\mathrm{T}_{n-1,k-1}p, \tag{1.37}$$

with the conventions $\mathrm{T}_{n,k}p := 0$ whenever $k > n$ or $k < 0$. Note that

$$\deg \mathrm{T}_{n,k}p \leq \deg p - k.$$

Theorem 1.6 (Leibniz Formula). *Let* $\mathbf{u} \in \mathcal{P}'$ *and* $p \in \mathcal{P}$. *Then*

$$\mathbf{D}^n(p\mathbf{u}) = \sum_{k=0}^{n} \mathrm{T}_{n,k}p\,\mathbf{D}^{n-k}\mathbf{S}^k\mathbf{u}, \quad n \in \mathbb{N}_0. \tag{1.38}$$

Proof. The proof is by induction on n. Of course, (1.38) holds for $n = 0$. Suppose now that (1.38) holds for a fixed nonnegative integer n. Then

$$\mathbf{D}^{n+1}(f\mathbf{u}) = \mathbf{D}(\mathbf{D}^n(f\mathbf{u})) = \sum_{k=0}^{n}\mathbf{D}(\mathrm{T}_{n,k}f\mathbf{D}^{n-k}\mathbf{S}^k\mathbf{u}). \tag{1.39}$$

From (1.35), we have

$$\mathbf{S}\mathbf{D}^{n-k}\mathbf{S}^k\mathbf{u} = \frac{1}{\alpha_{n+1-k}}\Big(\alpha\mathbf{D}^{n-k}\mathbf{S}^{k+1}\mathbf{u} - \gamma_{n-k}\mathrm{U}_1\mathbf{D}^{n+1-k}\mathbf{S}^k\mathbf{u}\Big), \tag{1.40}$$

and so

$$\begin{aligned}\mathbf{D}(\mathrm{T}_{n,k}p\mathbf{D}^{n-k}\mathbf{S}^k\mathbf{u}) &= (\mathrm{S}\mathrm{T}_{n,k}p - \alpha^{-1}\mathrm{U}_1\mathrm{D}\mathrm{T}_{n,k}p)\mathbf{D}^{n+1-k}\mathbf{S}^k\mathbf{u} \\ &\quad + \alpha^{-1}\mathrm{D}\mathrm{T}_{n,k}p\mathbf{S}\mathbf{D}^{n-k}\mathbf{S}^k\mathbf{u} \\ &= \left(\mathrm{S}\mathrm{T}_{n,k}p - \frac{\gamma_{n+1-k}}{\alpha_{n+1-k}}\mathrm{U}_1\mathrm{D}\mathrm{T}_{n,k}p\right)\mathbf{D}^{n+1-k}\mathbf{S}^k\mathbf{u} \\ &\quad + \frac{\mathrm{D}\mathrm{T}_{n,k}p}{\alpha_{n+1-k}}\mathbf{D}^{n-k}\mathbf{S}^{k+1}\mathbf{u} \\ &= \left(\mathrm{T}_{n+1,k}p - \frac{\mathrm{D}\mathrm{T}_{n,k-1}p}{\alpha_{n+2-k}}\right)\mathbf{D}^{n+1-k}\mathbf{S}^k\mathbf{u} \\ &\quad + \frac{\mathrm{D}\mathrm{T}_{n,k}p}{\alpha_{n+1-k}}\mathbf{D}^{n-k}\mathbf{S}^{k+1}\mathbf{u},\end{aligned}$$

which follows using successively (1.26), (1.40), (1.12), and (1.42). Substituting the last expression on the right-hand side of (1.39) and applying the method of telescoping sums, we obtain

$$\mathbf{D}^{n+1}(p\mathbf{u}) = \sum_{k=0}^{n}\mathrm{T}_{n+1,k}p\mathbf{D}^{n+1-k}\mathbf{S}^k\mathbf{u} + \frac{\mathrm{D}\mathrm{T}_{n,n}p}{\alpha_1}\mathbf{S}^{n+1}\mathbf{u} - \frac{\mathrm{D}\mathrm{T}_{n,-1}p}{\alpha_{n+2}}\mathbf{D}^{n+1}\mathbf{u}.$$

Finally, since $\mathrm{T}_{n,-1}p = 0$ and

$$\frac{1}{\alpha_1}\mathrm{D}\mathrm{T}_{n,n}p = \mathrm{T}_{n+1,n+1}p,$$

we obtain (1.38) with n replaced by $n+1$, and the theorem is proved. □

Corollary 1.1. *Let* $\mathbf{u} \in \mathcal{P}'$ *and* $\phi \in \mathcal{P}$. *Then*

$$D^n(\phi\mathbf{u}) = \sum_{k=0}^{n}\binom{n}{k}\phi^{(k)}\,D^{n-k}\mathbf{u}, \quad n \in \mathbb{N}_0.$$

Proof. Taking $x(s) = \mathfrak{c}_6$, we have

$$\mathrm{T}_{n,k}\phi = \mathrm{T}_{n-1,k}\phi + \mathrm{D}\mathrm{T}_{n-1,k-1}\phi, \tag{1.41}$$

for each $n \geq 1$ and $0 \leq k \leq n$. The proof is by induction on k. From (1.41), we get $\mathrm{T}_{n,0}\phi = \mathrm{T}_{n-1,0}\phi$, and so $\mathrm{T}_{n,0}\phi = \phi$. Suppose that

$$\mathrm{T}_{n,k-1}\phi = \binom{n}{k-1}\phi^{(k-1)}. \tag{1.42}$$

From (1.41), we get

$$\begin{aligned}\mathrm{T}_{n,k}\phi &= \mathrm{T}_{n,k-1}\phi + \binom{n-1}{k-1}\mathrm{D}\phi^{(k-1)} = \mathrm{T}_{n,k-1}\phi + \binom{n-1}{k-1}\phi^{(k)} \\ &= \left(\binom{n-1}{k} + \binom{n-1}{k-1}\right)\phi^{(k)} = \binom{n}{k}\phi^{(k)},\end{aligned}$$

and the result follows. □

1.5 The Formal Stieltjes Series

Let Δ' be the vector space of the formal series in the variable z with coefficients in $\mathbb{C}$, i.e.

$$\Delta' := \left\{ \sum_{n=0}^{\infty} c_n z^n \;\middle|\; c_n \in \mathbb{C}, n \in \mathbb{N}_0 \right\}.$$

In Δ' the operations of addition, multiplication and scalar multiplication are defined in the usual way. Endowing Δ' with the family of seminorms $\rho_n : \Delta' \to [0, +\infty)$, where

$$\rho_n\left(\sum_{n=0}^{\infty} c_n z^n\right) := \max_{0 \leq j \leq n} |c_j|, \quad n \in \mathbb{N}_0,$$

Δ' becomes a metrizable LCS. This space can be identified with $\mathcal{P}'$.

Theorem 1.7. *The operator $F : \mathcal{P}' \to \Delta'$ given by*

$$\mathbf{u} \in \mathcal{P}' \quad \mapsto \quad F(\mathbf{u}) := \sum_{n=0}^{\infty} u_n z^n$$

is a topological isomorphism.

Proof. Clearly, F is linear and bijective. Moreover, we get

$$\rho_n(F(\mathbf{u})) = \max_{0 \leq j \leq n} |u_j| = |\mathbf{u}|_n, \quad n \in \mathbb{N}_0.$$

Therefore, since the family of seminorms $\{|\cdot|_n : n \in \mathbb{N}_0\}$ generates the topology in $\mathcal{P}'$ (see Theorem 1.2), we deduce, for each sequence $\{\mathbf{u}_j\}_{j\geq 0}$ in $\mathcal{P}'$,

$$\begin{aligned} F(\mathbf{u}_j) \to 0 \ \ (\text{in } \Delta') \quad &\text{iff} \quad \rho_n(F(\mathbf{u}_j)) \to 0 \ \text{ for each } n \in \mathbb{N}_0, \\ &\text{iff} \quad |\mathbf{u}_j|_n \to 0 \ \text{ for each } n \in \mathbb{N}_0, \\ &\text{iff} \quad \mathbf{u}_j \to \mathbf{0}. \end{aligned}$$

Thus, F is bicontinuous, which proves the theorem. □

The isomorphism F allows us to transfer the algebraic structure from Δ' into $\mathcal{P}'$. This fact is accomplished through the formal Stieltjes series.

Definition 1.6. *Let $\mathbf{u} \in \mathcal{P}'$. The formal Stieltjes series associated with $\mathbf{u}$ is*

$$S_{\mathbf{u}}(z) := -\sum_{n=0}^{\infty} \frac{u_n}{z^{n+1}}.$$

Note that $S_{\mathbf{u}}(z)$ gives a representation for the sequence of moments, $\{u_n\}_{n\geq 0}$, of $\mathbf{u}$. The formal Stieltjes series is an important tool in the theory of OP, allowing us to state characterization theorems concerning certain important classes of OP, e.g., classical OP, semiclassical OP, and Laguerre-Hahn OP. The formal Stieltjes series $S_{\mathbf{u}}(z)$ and its formal derivative,

$$S'_{\mathbf{u}}(z) := \sum_{n=0}^{\infty} \frac{(n+1)u_n}{z^{n+2}},$$

become tools of major importance in the study of these classes of OP.

Exercises

1. Show that in any infinite dimensional normed space there are linear functionals which are not continuous. As a consequence, X being a normed space, there holds:

$$\begin{cases} X' = X^*, & \dim X < \infty; \\ X' \subsetneq X^*, & \dim X = \infty. \end{cases}$$

2. Let $\mathbb{P} := \mathbb{K}[x]$ be the set of all polynomials with coefficients in $\mathbb{K}$ ($= \mathbb{R}$

or $\mathbb{C}$). Prove that the mapping $\|\cdot\|:\mathbb{P}\to[0,+\infty)$ defined by

$$\|f\| := \max_{0\le k\le \deg f} |a_k|\,, \quad f(x)=\sum_{k=0}^{\deg f} a_k x^k \in \mathbb{K}[x],$$

is a norm in $\mathbb{P}$, but with this norm $\mathbb{P}$ is not a complete (Banach) space.

3. Prove that $\mathcal{P}'$ is a Fréchet space.

4. Show that the dual basis $\{\mathbf{a}_n\}_{n\ge0}$ corresponding to the simple set $\{x^n\}_{n\ge0}$ is given by

$$\mathbf{a}_n := \frac{(-1)^n}{n!}\delta^{(n)},$$

where $\delta^{(n)}$ is the (distributional) derivative of order n of the Dirac functional $\delta=\delta_0$. Conclude that each functional $\mathbf{u}\in\mathcal{P}'$ admits the representation

$$\mathbf{u}=\sum_{n=0}^{\infty}(-1)^n\frac{u_n}{n!}\delta^{(n)},$$

in the sense of the weak dual topology in $\mathcal{P}'$.

5. Prove the following identities:

$$\mathrm{S}(fg)=(\mathrm{D}f)(\mathrm{D}g)\mathtt{U}_2+(\mathrm{S}f)(\mathrm{S}g), \tag{1.43}$$

$$\mathrm{SD}f=\alpha\mathrm{DS}f-\mathrm{D}(\mathtt{U}_1\mathrm{D}f),$$

$$f\mathrm{S}g=\mathrm{S}((\mathrm{S}f-\alpha^{-1}\mathtt{U}_1\,\mathrm{D}f)g)-\alpha^{-1}\mathtt{U}_2\mathrm{D}(g\mathrm{D}f), \tag{1.44}$$

$$f\mathrm{D}g=\mathrm{D}((\mathrm{S}f-\alpha^{-1}\mathtt{U}_1\,\mathrm{D}f)g)-\alpha^{-1}\mathrm{S}(g\mathrm{D}f), \tag{1.45}$$

$$\mathbf{DSu}=\alpha\mathbf{SDu}+\mathbf{D}(\mathtt{U}_1\mathbf{Du}).$$

6. Prove Lemma 1.2.

7. Prove the properties listed in Proposition 1.2 and Proposition 1.4.

Chapter 2

Orthogonal Polynomial Sequences

In this chapter, we introduce the definition of orthogonality and some basic properties of orthogonal polynomial sequences. Moreover, we give necessary and sufficient conditions that guarantee the existence of an orthogonal polynomial sequence. In addition, the Favard theorem and the Christoffel-Darboux identity are proven.

2.1 Orthogonal Polynomial Sequences

Definition 2.1. *Let* $\mathbf{u} \in \mathcal{P}'$ *and let* $\{P_n\}_{n\geq 0}$ *be a sequence in* $\mathcal{P}$.

(i) $\{P_n\}_{n\geq 0}$ *is called an orthogonal polynomial sequence (OPS) with respect to* $\mathbf{u}$ *if* $\{P_n\}_{n\geq 0}$ *is a simple set and there exists a sequence of complex numbers* $\{h_n\}_{n\geq 0}$, *with* $h_n \in \mathbb{C} \setminus \{0\}$, *such that*

$$\langle P_m \mathbf{u}, P_n \rangle = h_n \delta_{m,n}, \quad m \in \mathbb{N}_0;$$

(ii) $\mathbf{u}$ *is called regular or quasi-definite if there exists an OPS with respect to* $\mathbf{u}$.

Whenever $\mathbf{u}$ is regular and $\{P_n\}_{n\geq 0}$ is an OPS with respect to $\mathbf{u}$, we will use phrases such as "$\{P_n\}_{n\geq 0}$ *is an OPS with respect to* $\mathbf{u}$", or "$\{P_n\}_{n\geq 0}$ *is an OPS associated with* $\mathbf{u}$". If $\mathbf{u}$ is regular and $\{P_n\}_{n\geq 0}$ is an associated OPS, then every polynomial admits a Fourier-type expansion in terms of a finite subset of $\{P_n\}_{n\geq 0}$.

Theorem 2.1. *Let* $\mathbf{u} \in \mathcal{P}'$ *be regular and let* $\{P_n\}_{n\geq 0}$ *be an OPS with*

DOI: 10.1201/9781003432067-2

respect to $\mathbf{u}$. *Let* π_k *be a polynomial of degree* k. *Then*

$$\pi_k(x) = \sum_{j=0}^{k} c_{k,j} P_j(x), \quad c_{k,j} := \frac{\langle \mathbf{u}, \pi_k P_j \rangle}{\langle \mathbf{u}, P_j^2 \rangle}.$$

Proof. Since $\{P_j\}_{j\geq 0}$ is a simple set in $\mathcal{P}$, $\{P_j\}_{j=0}^{k}$ is a basis of $\mathcal{P}_k$. Therefore, since $\pi_k \in \mathcal{P}_k$,

$$\pi_k(x) = \sum_{j=0}^{k} c_{k,j} P_j(x), \quad c_{k,j} \in \mathbb{C}.$$

Multiplying both sides of this equality by P_ℓ, $0 \leq \ell \leq k$, and taking the action of the functional $\mathbf{u}$ on both sides of the resulting equality, we deduce

$$\langle \mathbf{u}, \pi_k P_\ell \rangle = \sum_{j=0}^{k} c_{k,j} \langle \mathbf{u}, P_j P_\ell \rangle = c_{k,\ell} \langle \mathbf{u}, P_\ell^2 \rangle,$$

and the desired conclusion follows. □

Theorem 2.2. *Let* $\mathbf{u} \in \mathcal{P}'$ *and let* $\{P_n\}_{n\geq 0}$ *be a simple set in* $\mathcal{P}$. *Then the following conditions are equivalent:*

(i) $\{P_n\}_{n\geq 0}$ *is an OPS with respect to* $\mathbf{u}$;

(ii) *for each* $n \in \mathbb{N}_0$ *and* $\pi \in \mathcal{P}_n \setminus \{0\}$, *there is* $h_n = h_n(\pi) \in \mathbb{C} \setminus \{0\}$, *such that*

$$\langle \pi \mathbf{u}, P_n \rangle = h_n \delta_{m,n}, \quad m := \deg \pi.$$

(iii) *for each* $n \in \mathbb{N}_0$, *there exists* $h_n \in \mathbb{C} \setminus \{0\}$ *such that*

$$\langle x^m \mathbf{u}, P_n \rangle = h_n \delta_{m,n}, \quad m = 0, 1, \ldots, n.$$

Proof. Assume that (i) holds. Fix $n \in \mathbb{N}_0$ and let $\pi \in \mathcal{P}_n$. Setting $m := \deg \pi$, by Theorem 2.1, we know that there exist complex numbers $c_{m,j}$ such that

$$\pi(x) = \sum_{j=0}^{m} c_{m,j} P_j(x).$$

Clearly, $c_{m,m} \neq 0$, because $\pi \neq 0$, $\deg \pi = m$, and $\{P_j\}_{j\geq 0}$ is a simple set in $\mathcal{P}$. Hence

$$\langle \mathbf{u}, \pi P_n \rangle = \sum_{j=0}^{m} c_{m,j} \langle \mathbf{u}, P_j P_n \rangle = \begin{cases} 0 & \text{if } m < n, \\ c_{n,n} \langle \mathbf{u}, P_n^2 \rangle \neq 0 & \text{if } m = n, \end{cases}$$

and so (i)⇒(ii), where $h_n := c_{n,n}\langle \mathbf{u}, P_n^2\rangle$. Taking $\pi(x) := x^m$ in (ii), it is clear that (ii)⇒(iii). Finally, assume that (iii) holds. Fix $j, n \in \mathbb{N}_0$ and, without loss of generality, assume that $j \leq n$. Since $\{P_k\}_{k\geq 0}$ is a simple set, there exist complex numbers $c_{j,m}$, with $c_{j,j} \neq 0$, such that $P_j(x) = \sum_{m=0}^{j} c_{j,m}x^m$. Thus

$$\langle \mathbf{u}, P_j P_n\rangle = \sum_{m=0}^{j} c_{j,m}\langle \mathbf{u}, x^m P_n\rangle = \sum_{m=0}^{j} c_{j,m} h_n \delta_{m,n} = \widetilde{h}_n \delta_{j,n},$$

where $\widetilde{h}_n := c_{n,n}h_n \neq 0$, and so (iii)⇒(i), which completes the proof. □

The next result states that, up to normalization, there exists only one OPS associated with a given regular functional.

Theorem 2.3. *Let $\mathbf{u} \in \mathcal{P}'$ be regular and let $\{P_n\}_{n\geq 0}$ and $\{Q_n\}_{n\geq 0}$ be two OPS with respect to $\mathbf{u}$. Then, there exists a sequence $\{c_n\}_{n\geq 0}$ with $c_n \in \mathbb{C} \setminus \{0\}$ such that*

$$Q_n(x) = c_n P_n(x).$$

Proof. Fix $k \in \mathbb{N}$. Since $\{Q_n\}_{n\geq 0}$ is an OPS with respect to $\mathbf{u}$,

$$\langle \mathbf{u}, Q_k P_j\rangle = 0 \quad \text{if} \quad j < k.$$

Thus, by Theorem 2.1, taking $\pi_k(x) = Q_k(x)$, we obtain $Q_k(x) = c_{k,k}P_k(x)$, where

$$c_{k,k} = \frac{\langle \mathbf{u}, P_k Q_k\rangle}{\langle \mathbf{u}, P_k^2\rangle} \neq 0,$$

and the result follows by taking $c_k := c_{k,k}$. □

Theorem 2.3 implies that an OPS is uniquely determined if it satisfies a condition that fixes the leading coefficient of each P_n (i.e., the coefficient of x^n). In particular, if $\{P_n\}_{n\geq 0}$ is an OPS and the leading coefficient of each P_n is 1, we say that $\{P_n\}_{n\geq 0}$ is the monic OPS. Of course, given a not necessarily monic OPS, $\{P_n\}_{n\geq 0}$, with

$$P_n(x) = k_n x^n + (\text{lower degree terms}),$$

the elements of the corresponding monic sequence, $\{\widehat{P}_n\}_{n\geq 0}$, are given by

$$\widehat{P}_n(x) := k_n^{-1} P_n(x).$$

On the other hand, if $\{P_n\}_{n\geq 0}$ is an OPS with respect to $\mathbf{u}$ and

$$\langle \mathbf{u}, P_n^2\rangle = 1,$$

we say that $\{P_n\}_{n\geq 0}$ is an orthonormal polynomial sequence. In general, $\{P_n\}_{n\geq 0}$ being an OPS with respect to $\mathbf{u}$ (not necessarily orthonormal), the sequence $\{p_n\}_{n\geq 0}$, where

$$p_n(x) := \langle \mathbf{u}, P_n^2\rangle^{-1/2} P_n(x),$$

is orthonormal with respect to $\mathbf{u}$. Here the square root needs not be real, but, as noticed above, $p_n(x)$ may be uniquely determined by requiring an additional condition on its leading coefficient (e.g., that its leading coefficient be positive). Finally, we notice the following obvious fact: *If $\{P_n\}_{n\geq 0}$ is an OPS with respect to the functional $\mathbf{u} \in \mathcal{P}'$, then $\{P_n\}_{n\geq 0}$ is also an OPS with respect to the functional $c\,\mathbf{u}$, for every constant $c \in \mathbb{C}\setminus\{0\}$.*

2.2 Existence of Orthogonal Polynomial Sequences

In this section, we analyze the question of whether a given functional $\mathbf{u} \in \mathcal{P}'$ is regular, i.e., we ask for necessary and sufficient conditions that guarantee the existence of an OPS with respect to $\mathbf{u}$. To answer this question, we introduce the so-called Hankel determinants. Denoting, as usual, the moments of $\mathbf{u}$ by $u_j := \langle \mathbf{u}, x^j\rangle$, $j \in \mathbb{N}_0$, we define the associated Hankel determinant $H_n \equiv H_n(\mathbf{u})$ as

$$H_n := \det\left\{[u_{i+j}]_{i,j=0}^{n}\right\} = \begin{vmatrix} u_0 & u_1 & \cdots & u_{n-1} & u_n \\ u_1 & u_2 & \cdots & u_n & u_{n+1} \\ \vdots & \vdots & \ddots & \vdots & \vdots \\ u_{n-1} & u_n & \cdots & u_{2n-2} & u_{2n-1} \\ u_n & u_{n+1} & \cdots & u_{2n-1} & u_{2n} \end{vmatrix}, \quad n \in \mathbb{N}_0. \tag{2.1}$$

Notice that H_n is a determinant of order $n+1$. It is also useful to set

$$H_{-1} := 1. \tag{2.2}$$

Theorem 2.4. *The functional $\mathbf{u} \in \mathcal{P}'$ is regular if and only if, for each $n \in \mathbb{N}_0$,*

$$H_n \neq 0. \tag{2.3}$$

Under such conditions, the monic OPS, $\{P_n\}_{n\geq 0}$, with respect to $\mathbf{u}$ is *given by* $P_0(x) = 1$ *and*

$$P_n(x) = \frac{1}{H_{n-1}} \begin{vmatrix} u_0 & u_1 & \cdots & u_{n-1} & u_n \\ u_1 & u_2 & \cdots & u_n & u_{n+1} \\ \vdots & \vdots & \ddots & \vdots & \vdots \\ u_{n-1} & u_n & \cdots & u_{2n-2} & u_{2n-1} \\ 1 & x & \cdots & x^{n-1} & x^n \end{vmatrix}, \quad n \in \mathbb{N}. \tag{2.4}$$

Proof. Suppose that $\mathbf{u}$ is regular. Let $\{P_n\}_{n\geq 0}$ be an OPS with respect to $\mathbf{u}$. For fixed $n \in \mathbb{N}_0$, there exist $c_{n,0}, c_{n,1}, \ldots, c_{n,n} \in \mathbb{C}$ such that

$$P_n(x) = \sum_{k=0}^{n} c_{n,k} x^k. \tag{2.5}$$

By Theorem 2.2, there exists $h_n \in \mathbb{C} \setminus \{0\}$ such that

$$h_n \delta_{m,n} = \langle \mathbf{u}, x^m P_n \rangle = \sum_{k=0}^{n} c_{n,k} u_{k+m}, \quad m = 0, 1, \ldots, n. \tag{2.6}$$

This may be written in matrix form as

$$\begin{bmatrix} u_0 & u_1 & \cdots & u_{n-1} & u_n \\ u_1 & u_2 & \cdots & u_n & u_{n+1} \\ \vdots & \vdots & \ddots & \vdots & \vdots \\ u_{n-1} & u_n & \cdots & u_{2n-2} & u_{2n-1} \\ u_n & u_{n+1} & \cdots & u_{2n-1} & u_{2n} \end{bmatrix} \begin{bmatrix} c_{n,0} \\ c_{n,1} \\ \vdots \\ c_{n,n-1} \\ c_{n,n} \end{bmatrix} = \begin{bmatrix} 0 \\ 0 \\ \vdots \\ 0 \\ h_n \end{bmatrix}. \tag{2.7}$$

Since the sequence $\{h_n\}_{n\geq 0}$ in (2.6) uniquely determines $\{P_n\}_{n\geq 0}$, it follows that the system (2.7), where the coefficients $c_{n,0}, c_{n,1}, \ldots, c_{n,n}$ of P_n are the unknowns, has a unique solution. Hence $H_n \neq 0$, because H_n is the determinant of such a system. Conversely, suppose that $H_n \neq 0$ for all $n \in \mathbb{N}_0$. Then, for any fixed $n \in \mathbb{N}_0$, to each constant $h_n \in \mathbb{C} \setminus \{0\}$ corresponds a unique vector $[c_{n,0}\, c_{n,1}\, \ldots\, c_{n,n}]^T$ solution of the system (2.7). Using the components of this vector, we may define a polynomial $P_n(x)$ through the expression (2.5). This polynomial fulfills (2.6), because (2.6) and (2.7) are equivalent. To conclude that $\{P_n\}_{n\geq 0}$ is an OPS with respect to $\mathbf{u}$, it remains to prove that it is a simple set, i.e., $\deg P_n = n$ for all n. Indeed, solving (2.7) for $c_{n,n}$ by

Crammer's rule, and taking into account the hypothesis $H_n \neq 0$, we obtain

$$c_{n,n} = \frac{h_n H_{n-1}}{H_n}, \tag{2.8}$$

hence $c_{n,n} \neq 0$, which proves that, indeed, $\deg P_n = n$ for all n. It remains to prove (2.4):

First proof of (2.4): For each fixed $n \in \mathbb{N}$, $P_n(x)$ may be written as in (2.5), where $c_{n,0}, c_{n,1}, \ldots, c_{n,n-1} \in \mathbb{C}$, and $c_{n,n} = 1$. As explained, for each $m \in \{0, 1, \cdots, n-1\}$, we deduce

$$0 = \langle \mathbf{u}, x^m P_n \rangle = \sum_{k=0}^{n} c_{n,k} u_{k+m}.$$

From this, and taking into account that $c_{n,n} = 1$, we obtain the following system of n equations in the n unknowns $c_{n,0}, c_{n,1}, \ldots, c_{n,n-1}$:

$$\begin{bmatrix} u_0 & u_1 & \cdots & u_{n-1} \\ u_1 & u_2 & \cdots & u_n \\ \vdots & \vdots & \ddots & \vdots \\ u_{n-1} & u_n & \cdots & u_{2n-2} \end{bmatrix} \begin{bmatrix} c_{n,0} \\ c_{n,1} \\ \vdots \\ c_{n,n-1} \end{bmatrix} = \begin{bmatrix} -u_n \\ -u_{n+1} \\ \vdots \\ -u_{2n-1} \end{bmatrix}.$$

The determinant of this system is $H_{n-1} \neq 0$. Solving by Crammer's rule, we obtain

$$c_{n,k} = \frac{1}{H_{n-1}} \begin{vmatrix} u_0 & \cdots & u_{k-1} & -u_n & u_{k+1} & \cdots & u_{n-1} \\ u_1 & \cdots & u_k & -u_{n+1} & u_{k+2} & \cdots & u_n \\ \vdots & \ddots & \vdots & \vdots & \vdots & \ddots & \vdots \\ u_{n-1} & \cdots & u_{n+k-2} & -u_{2n-1} & u_{n+k} & \cdots & u_{2n-2} \end{vmatrix}$$

for each $k = 0, 1, \ldots, n-1$. Performing elementary operations on the columns of this determinant, by moving successively the $(k+1)$th column to its right (so that $n-k-1$ permutations on columns must be done), we deduce

$$c_{n,k} = \frac{(-1)^{n-k}}{H_{n-1}} \begin{vmatrix} u_0 & \cdots & u_{k-1} & u_{k+1} & \cdots & u_{n-1} & u_n \\ u_1 & \cdots & u_k & u_{k+2} & \cdots & u_n & u_{n+1} \\ \vdots & \ddots & \vdots & \vdots & \ddots & \vdots & \vdots \\ u_{n-1} & \cdots & u_{n+k-2} & u_{n+k} & \cdots & u_{2n-2} & u_{2n-1} \end{vmatrix} \tag{2.9}$$

for each $k = 0, 1, \ldots, n-1$. Clearly, (2.9) is also true for $k = n$, since in that case the right-hand side of (2.9) reduces to 1. Therefore, substituting (2.9) into (2.5), we obtain

$$P_n(x)$$

$$= \frac{1}{H_{n-1}} \sum_{k=0}^{n} (-1)^{n-k} \begin{vmatrix} u_0 & \cdots & u_{k-1} & u_{k+1} & \cdots & u_{n-1} & u_n \\ u_1 & \cdots & u_k & u_{k+2} & \cdots & u_n & u_{n+1} \\ \vdots & \ddots & \vdots & \vdots & \ddots & \vdots & \vdots \\ u_{n-1} & \cdots & u_{n+k-2} & u_{n+k} & \cdots & u_{2n-2} & u_{2n-1} \end{vmatrix} x^k.$$

Thus, the formula (2.4) follows by Laplace's theorem, developing the determinant in the right-hand side of (2.4) along its last row.

Second proof of (2.4): Let $Q_n(x)$ be the (monic) polynomial of degree n defined by the right-hand side of (2.4). If $m < n$ then, clearly, $\langle \mathbf{u}, x^m Q_n \rangle = 0$, because $\langle \mathbf{u}, x^m Q_n \rangle$ becomes a determinant whose $m+1$ row and $n+1$ row are equal. If $m = n$, then we simply notice that

$$\langle \mathbf{u}, x^n Q_n \rangle = \frac{H_n}{H_{n-1}} \neq 0.$$

Thus, by Theorem 2.2, $\{Q_n\}_{n\geq 0}$ is an OPS with respect to $\mathbf{u}$, and since each $Q_n(x)$ is a monic polynomial, (2.4) follows. □

Corollary 2.1. *Let $\{P_n\}_{n\geq 0}$ be an OPS with respect to $\mathbf{u} \in \mathcal{P}'$ and let π_n be a polynomial of degree n. Denote by k_n and a_n the leading coefficients of P_n and π_n, respectively, so that*

$$P_n(x) = k_n x^n + \text{(lower degree terms)},$$
$$\pi_n(x) = a_n x^n + \text{(lower degree terms)}.$$

Then

$$\langle \mathbf{u}, \pi_n P_n \rangle = a_n \langle \mathbf{u}, x^n P_n \rangle = a_n\, k_n \frac{H_n}{H_{n-1}}. \tag{2.10}$$

Proof. Writing $\pi_n(x) = a_n x^n + \pi_{n-1}(x)$, with $\pi_{n-1} \in \mathcal{P}_{n-1}$, and taking into account Theorem 2.2, we get

$$\langle \mathbf{u}, \pi_n P_n \rangle = a_n \langle \mathbf{u}, x^n P_n \rangle + \langle \mathbf{u}, \pi_{n-1} P_n \rangle$$
$$= a_n \langle \mathbf{u}, x^n P_n \rangle = a_n h_n = a_n\, k_n \frac{H_n}{H_{n-1}},$$

where the last equality follows from (2.8), noticing that $c_{n,n} = k_n$. □

2.3 The Positive Definite Case

In many important cases, the functional $\mathbf{u} \in \mathcal{P}'$ with respect to which the polynomials are orthogonal admits an integral representation involving a weight function, or, in the most general situation, a positive Borel measure, μ, whose support is an infinite subset of $\mathbb{R}$, and with finite moments of all orders, so that

$$\langle \mathbf{u}, p \rangle = \int p(x)\,\mathrm{d}\mu(x), \quad p \in \mathcal{P}. \tag{2.11}$$

One easily verifies that, under such conditions, the property

$$\langle \mathbf{u}, p \rangle > 0 \tag{2.12}$$

holds for each polynomial $p \in \mathcal{P}$ which is nonzero (i.e., it does not vanish identically) and nonnegative for all $x \in \mathbb{R}$. It turns out that this property characterizes functionals $\mathbf{u} \in \mathcal{P}'$ such that an integral representation as (2.11) holds, under the conditions described above. This "equivalence" between (2.11) and (2.12) is a nontrivial fact, and it will be proved later. We start the study of such functionals by introducing the following definition.

Definition 2.2. *A functional $\mathbf{u} \in \mathcal{P}'$ is called positive definite if, for each polynomial $p(x)$ which is nonzero and nonnegative for all real x, the condition*

$$\langle \mathbf{u}, p \rangle > 0 \tag{2.13}$$

holds.

Next, we state some basic properties of positive definite linear functionals in $\mathcal{P}$.

Theorem 2.5. *Let $\mathbf{u} \in \mathcal{P}'$ be positive definite. Then, the moments $u_n := \langle \mathbf{u}, x^n \rangle$, $n \in \mathbb{N}_0$, are real numbers. More precisely, the following holds:*

$$u_{2n} > 0, \quad u_{2n+1} \in \mathbb{R}. \tag{2.14}$$

Proof. On the first hand, since $\mathbf{u}$ is positive definite and $x^{2n} \geq 0$ for $x \in \mathbb{R}$, then

$$u_{2n} = \langle \mathbf{u}, x^{2n} \rangle > 0.$$

On the other hand, using again the positive definiteness of $\mathbf{u}$ and Newton's binomial formula, we may write

$$0 < \langle \mathbf{u}, (1+x)^{2n} \rangle = \sum_{k=0}^{2n} \binom{2n}{k} u_k,$$

hence it follows by induction that u_{2n+1} is a real number, and the theorem is proved. □

Given a positive definite functional defined in $\mathcal{P}$, a corresponding OPS can be constructed using a step-by-step method known as the Gram-Schmidt process.

Theorem 2.6 (Gram-Schmidt Process). *Let $\mathbf{u} \in \mathcal{P}'$ be positive definite. Define a sequence of polynomials $\{p_n\}_{n\geq 0}$ as follows:*

$$p_n(x) := \langle \mathbf{u}, P_n^2 \rangle^{-1/2} P_n(x), \tag{2.15}$$

where $\{P_n\}_{n\geq 0}$ is a simple set of monic polynomials, constructed step-by-step as

$$P_n(x) := x^n - \sum_{k=0}^{n-1} \langle \mathbf{u}, x^n p_k \rangle \, p_k(x). \tag{2.16}$$

Then, $\{p_n\}_{n\geq 0}$ is orthonormal with respect to $\mathbf{u}$, $p_n(x)$ being a real polynomial (i.e., with real coefficients). Moreover, $\{P_n\}_{n\geq 0}$ is the corresponding monic OPS.

Proof. To state the result we prove that each $P_n(x)$ is a real polynomial and

$$\langle \mathbf{u}, p_n^2 \rangle = 1, \quad \langle \mathbf{u}, p_j p_{n+1} \rangle = 0, \quad j = 0, \ldots, n. \tag{2.17}$$

Since $\mathbf{u}$ is positive definite, $\langle \mathbf{u}, P_n^2 \rangle > 0$, and so $p_n(x)$ is also a real. The proof of (2.17) is by induction on n. For $n = 0$, we have

$$P_0(x) = 1, \quad p_0(x) = u_0^{-1/2},$$

and so $p_0(x)$ is real—notice that, by Theorem 2.5, conditions (2.14) hold—and

$$\langle \mathbf{u}, p_0^2 \rangle = u_0^{-1} \langle \mathbf{u}, 1 \rangle = u_0^{-1} u_0 = 1.$$

Now, we compute

$$P_1(x) = x - \langle \mathbf{u}, x p_0 \rangle p_0(x) = x - \frac{u_1}{u_0},$$

hence $P_1(x)$ is a real polynomial, and since $\mathbf{u}$ is positive definite, we have $\langle \mathbf{u}, P_1^2 \rangle > 0$. Thus,

$$p_1(x) := \langle \mathbf{u}, P_1^2 \rangle^{-1/2} P_1(x)$$

is well defined, it is a real polynomial, and

$$\langle \mathbf{u}, p_0 p_1 \rangle = u_0^{-1/2} \langle \mathbf{u}, P_1^2 \rangle^{-1/2} \langle \mathbf{u}, x - u_1/u_0 \rangle = 0,$$

and we conclude that (2.17) holds for $n = 0$. Assume now that, for some $m \in \mathbb{N}_0$, the polynomials $P_1(x), \dots, P_{m+1}(x)$ are real, and (2.17) holds for all positive integers $n \leq m$. We need to prove that $P_{m+2}(x)$ is also a real polynomial and (2.17) remains true if n is replaced by $m+1$. Indeed, since $P_{m+1}(x)$ is real and $\mathbf{u}$ is positive definite, then $\langle \mathbf{u}, P_{m+1}^2 \rangle > 0$, and so

$$\langle \mathbf{u}, p_{m+1}^2 \rangle = \left\langle \mathbf{u}, \langle \mathbf{u}, P_{m+1}^2 \rangle^{-1} P_{m+1}^2 \right\rangle = 1.$$

Moreover, since $P_1(x), \dots, P_{m+1}(x)$ are real, $p_0(x), p_1(x), \dots, p_{m+1}(x)$ are real polynomials, and so $P_{m+2}(x)$ is also a real polynomial. Then, $\langle \mathbf{u}, P_{m+2}^2 \rangle > 0$, and so

$$p_{m+2}(x) := \langle \mathbf{u}, P_{m+2}^2 \rangle^{-1/2} P_{m+2}(x)$$

is well defined. Thus, for $j = 0, 1, \dots, m+1$,

$$\langle \mathbf{u}, p_j p_{m+2} \rangle = \langle \mathbf{u}, P_{m+2}^2 \rangle^{-1/2} \left(\langle \mathbf{u}, p_j x^{m+2} \rangle - \sum_{k=0}^{m+1} \langle \mathbf{u}, x^{m+2} p_k \rangle \langle \mathbf{u}, p_j p_k \rangle \right).$$

Since $\langle \mathbf{u}, p_j p_k \rangle = \delta_{j,k}$ if $j, k \in \{0, 1, \dots, m+1\}$, we deduce

$$\langle \mathbf{u}, p_j p_{m+2} \rangle = \langle \mathbf{u}, P_{m+2}^2 \rangle^{-1/2} \left(\langle \mathbf{u}, p_j x^{m+2} \rangle - \langle \mathbf{u}, x^{m+2} p_j \rangle \right) = 0,$$

and the result follows. □

Corollary 2.2. *Let $\mathbf{u} \in \mathcal{P}'$ be positive definite. Then, $\mathbf{u}$ is regular.*

Next, we state the connection between positive definite functionals defined in $\mathcal{P}$ and the Hankel determinants introduced in (2.1). We will need the following classical result characterizing non-negative polynomials.

Lemma 2.1. *Let $\pi(x)$ be a polynomial that is non-negative for all real x. Then, there are real polynomials $P(x)$ and $Q(x)$ such that*

$$\pi(x) = P^2(x) + Q^2(x). \tag{2.18}$$

Proof. Since $\pi(x) \geq 0$ for $x \in \mathbb{R}$, then π is a real polynomial (i.e., its coefficients are all real numbers) such that its real zeros have even multiplicity and its non-real zeros occur in conjugate pairs. Thus, we can write

$$\pi(x) = R^2(x) \prod_{k=1}^{m} (x - a_k + ib_k)(x - a_k - ib_k),$$

where R is a real polynomial and a_k and b_k are real numbers. Therefore, since we may write

$$\prod_{k=1}^{m} (x - a_k + ib_k) = A(x) + iB(x),$$

where A and B are real polynomials, we deduce

$$\pi(x) = R^2(x)(A(x) + iB(x))\overline{(A(x) + iB(x))} = R^2(x)\left(A^2(x) + B^2(x)\right).$$

Finally, the desired result follows by taking $P := RA$ and $Q := RB$. □

Theorem 2.7. *The functional* $\mathbf{u} \in \mathcal{P}'$ *is positive definite if and only if the following two conditions hold:*

(i) *The moments* $u_n := \langle \mathbf{u}, x^n \rangle$ *are real for each* $n \in \mathbb{N}_0$;

(ii) *the Hankel determinants* (2.1) *are all positive.*

Proof. Suppose that $\mathbf{u}$ is positive definite. Then, by Theorem 2.5, all the moments u_n are real. Moreover, by Theorem 2.6, a monic OPS $\{P_n\}_{n\geq 0}$ with respect to $\mathbf{u}$ exists, where each $P_n(x)$ is a real polynomial, and so $\langle \mathbf{u}, P_n^2 \rangle > 0$ for each $n \in \mathbb{N}_0$, because $\mathbf{u}$ is positive definite. Then, taking into account Corollary 2.1, we have

$$0 < \langle \mathbf{u}, P_n^2 \rangle = \frac{H_n}{H_{n-1}}.$$

Therefore, since $H_{-1} = 1$, it follows by induction that $H_n > 0$ for all $n \in \mathbb{N}_0$. Conversely, suppose that conditions (i) and (ii) hold. The condition (ii) and Theorem 2.4 ensure that $\mathbf{u}$ is regular, hence there exists a monic OPS $\{P_n\}_{n\geq 0}$ with respect to $\mathbf{u}$. Since $P_n(x)$ admits the representation (2.4), it follows from (i) and (ii) that each $P_n(x)$ is a real polynomial. Also, again by Corollary 2.1 and by (ii), we have

$$\langle \mathbf{u}, P_n^2 \rangle = \frac{H_n}{H_{n-1}} > 0.$$

Let $Q(x)$ be a nonzero real polynomial of degree m. Since each $P_n(x)$ is real, we may write $Q(x) = \sum_{j=0}^{m} a_j P_j(x)$, where $a_j \in \mathbb{R}$ for all j, with $a_m \neq 0$. Hence

$$\langle \mathbf{u}, Q^2 \rangle = \sum_{j=0}^{m} a_j^2 \langle \mathbf{u}, P_j^2 \rangle > 0,$$

and it follows from Lemma 2.1 that $\mathbf{u}$ is positive definite. □

Corollary 2.3. *Suppose that $\mathbf{u} \in \mathcal{P}'$ is regular and let $\{P_n\}_{n\geq 0}$ be the corresponding monic OPS. Assume further that for each n, $P_n(x)$ is real and*

$$\langle \mathbf{u}, P_n^2 \rangle > 0. \tag{2.19}$$

Then, $\mathbf{u}$ is positive definite.

Proof. The hypothesis allows us to proceed as in the last part of the proof of Theorem 2.7, in order to obtain $\langle \mathbf{u}, Q^2 \rangle > 0$ for every nonzero real polynomial Q, so that, by Lemma 2.1, $\mathbf{u}$ is positive definite. □

2.4 The Favard Theorem

One of the most important characterizations of OPS is the fact that any three consecutive polynomials are connected by a very simple relation. We begin by stating the following theorem.

Theorem 2.8. *Let $\mathbf{u} \in \mathcal{P}'$ be regular and let $\{P_n\}_{n\geq 0}$ be the corresponding monic OPS. Then, $\{P_n\}_{n\geq 0}$ satisfies the three-term recurrence relation*

$$P_{n+1}(x) = (x - \beta_n)P_n(x) - \gamma_n P_{n-1}(x), \tag{2.20}$$

with initial conditions

$$P_{-1}(x) = 0, \quad P_0(x) = 1, \tag{2.21}$$

where $\{\beta_n\}_{n\geq 0}$ and $\{\gamma_n\}_{n\geq 1}$ are sequences of complex numbers such that

$$\gamma_n \neq 0. \tag{2.22}$$

Moreover, if $\mathbf{u}$ is positive definite, then

$$\beta_{n-1} \in \mathbb{R}, \quad \gamma_n > 0. \tag{2.23}$$

Proof. Since $xP_n(x)$ is a polynomial of degree $n+1$, by Theorem 2.1 we get

$$xP_n(x) = \sum_{j=0}^{n+1} c_{n,j} P_j(x), \quad c_{n,j} := \frac{\langle \mathbf{u}, xP_nP_j\rangle}{\langle \mathbf{u}, P_j^2\rangle}, \quad j = 0, \dots, n+1.$$

Clearly, $\langle \mathbf{u}, xP_nP_j\rangle = 0$ if $j = 0, \dots, n-2$, and $c_{n,n+1} = 1$. Hence

$$xP_n(x) = P_{n+1}(x) + c_{n,n}P_n(x) + c_{n,n-1}P_{n-1}(x).$$

Therefore, we obtain (2.20), with

$$\beta_n := c_{n,n} = \frac{\langle \mathbf{u}, xP_n^2\rangle}{\langle \mathbf{u}, P_n^2\rangle},$$

and

$$\gamma_n := c_{n,n-1} = \frac{\langle \mathbf{u}, xP_nP_{n-1}\rangle}{\langle \mathbf{u}, P_{n-1}^2\rangle} = \frac{\langle \mathbf{u}, P_n^2\rangle}{\langle \mathbf{u}, P_{n-1}^2\rangle} \in \mathbb{C} \setminus \{0\}.$$

If $\mathbf{u}$ is a positive definite, then, by Theorem 2.6, each $P_n(x)$ is a real polynomial. Finally, it follows from the previous expressions for β_n and γ_n that conditions (2.23) hold, and the theorem is proved. □

Remark 2.1. *Since $P_{-1}(x) = 0$, then it does not matter how we define γ_0. Often we will make the useful choice $\gamma_0 := u_0$.*

Corollary 2.4. *Under the conditions of Theorem 2.8, the following holds:*

(i) *The β–parameters are given by*

$$\beta_n = \frac{\langle \mathbf{u}, xP_n^2\rangle}{\langle \mathbf{u}, P_n^2\rangle}; \tag{2.24}$$

(ii) *the γ–parameters are given by*

$$\gamma_n = \frac{\langle \mathbf{u}, P_n^2\rangle}{\langle \mathbf{u}, P_{n-1}^2\rangle} = \frac{H_{n-2}H_n}{H_{n-1}^2}; \tag{2.25}$$

(iii) *setting*

$$P_n(x) = x^n + f_nx^{n-1} + g_nx^{n-2} + (\text{lower degree terms}), \tag{2.26}$$

the coefficients f_n and g_n are given in terms of the β and γ-parameters by

$$f_n = -\sum_{j=0}^{n-1} \beta_j, \quad n \in \mathbb{N}, \tag{2.27}$$

and

$$g_n = \sum_{0 \leq i < j \leq n-1} \beta_i \beta_j - \sum_{k=1}^{n-1} \gamma_k, \quad n \in \mathbb{N} \setminus \{1\}. \tag{2.28}$$

Proof. The statements (i) and (ii) follow from the proof of Theorem 2.8 and taking into account (2.10), where

$$\langle \mathbf{u}, P_n^2 \rangle = \frac{H_n}{H_{n-1}}.$$

To prove (iii), substitute $P_n(x) = x^n + f_n x^{n-1} + g_n x^{n-2} + \cdots$ and the corresponding expressions for $P_{n+1}(x)$ and $P_{n-1}(x)$ in the recurrence relation (2.20), so that

$$\begin{aligned}
&x^{n+1} + f_{n+1}x^n + g_{n+1}x^{n-1} + \cdots \\
&\quad = (x - \beta_n)\left(x^n + f_n x^{n-1} + g_n x^{n-2} + \cdots\right) \\
&\qquad - \gamma_n\left(x^{n-1} + f_{n-1}x^{n-2} + \cdots\right) \\
&\quad = x^{n+1} + (f_n - \beta_n)x^n + (g_n - \beta_n f_n - \gamma_n)x^{n-1} + \cdots .
\end{aligned}$$

Therefore, by comparing coefficients, and defining $f_0 = g_1 = 0$, we obtain

$$f_{n+1} = f_n - \beta_n, \quad n \in \mathbb{N}_0,$$

$$g_{n+1} = g_n - \beta_n f_n - \gamma_n, \quad n \in \mathbb{N}.$$

Thus, (2.27) and (2.28) follow easily by induction (or by applying the telescoping property for sums). □

Remark 2.2. *Regarding Corollary 2.4, notice also the following relations (with the convention that the empty product equals one):*

$$\langle \mathbf{u}, P_n^2 \rangle = \frac{H_n}{H_{n-1}} = u_0 \prod_{j=1}^{n} \gamma_j. \tag{2.29}$$

Theorem 2.9 (Favard Theorem). *Let $\{\beta_n\}_{n\geq 0}$ and $\{\gamma_n\}_{n\geq 0}$ be two arbitrary sequences of complex numbers. Let $\{P_n\}_{n\geq 0}$ be a sequence of (monic) polynomials defined by the three-term recurrence relation*

$$P_{n+1}(x) = (x-\beta_n)P_n(x) - \gamma_n P_{n-1}(x), \tag{2.30}$$

with initial conditions

$$P_{-1}(x) = 0, \quad P_0(x) = 1. \tag{2.31}$$

Then there exists a unique functional $\mathbf{u} \in \mathcal{P}'$ such that

$$\langle \mathbf{u}, 1\rangle = u_0 := \gamma_0, \quad \langle \mathbf{u}, P_n P_m\rangle = 0 \quad \text{if} \quad n \neq m. \tag{2.32}$$

Moreover, $\mathbf{u}$ is regular and $\{P_n\}_{n\geq 0}$ is the corresponding monic OPS if and only if $\gamma_n \neq 0$ for each n. Furthermore, $\mathbf{u}$ is positive definite and $\{P_n\}_{n\geq 0}$ is the corresponding monic OPS if and only if $\beta_n \in \mathbb{R}$ and $\gamma_n > 0$ for each n.

Proof. Since $\{P_n\}_{n\geq 0}$, defined by (2.30), is clearly a simple set in $\mathcal{P}$ (so that it is an algebraic basis in $\mathcal{P}$), we may define a functional $\mathbf{u} \in \mathcal{P}'$ by

$$\langle \mathbf{u}, P_0\rangle = \langle \mathbf{u}, 1\rangle := \gamma_0, \quad \langle \mathbf{u}, P_n\rangle = 0. \tag{2.33}$$

Rewriting (2.30) as

$$xP_n(x) = P_{n+1}(x) + \beta_n P_n(x) + \gamma_n P_{n-1}(x) \tag{2.34}$$

yields

$$\langle \mathbf{u}, xP_n\rangle = \langle \mathbf{u}, P_{n+1}\rangle + \beta_n\langle \mathbf{u}, P_n\rangle + \gamma_n\langle \mathbf{u}, P_{n-1}\rangle, \quad n \in \mathbb{N}_0.$$

Therefore, by (2.33),

$$\langle \mathbf{u}, xP_n\rangle = 0, \quad n \in \mathbb{N}\setminus\{1\}. \tag{2.35}$$

Multiplying both sides of (2.34) by x and using (2.35), we find

$$\langle \mathbf{u}, x^2P_n\rangle = 0, \quad n \in \mathbb{N}\setminus\{1,2\}.$$

Continuing in this manner, we deduce

$$\langle \mathbf{u}, x^kP_n\rangle = 0, \quad k = 0,\dots,n. \tag{2.36}$$

Therefore, if $m \neq n$, say $m < n$, then by writing

$$P_m(x) = \sum_{k=0}^{m} a_{m,k}x^k, \quad a_{m,k} \in \mathbb{C},$$

we obtain

$$\langle \mathbf{u}, P_m P_n \rangle = \sum_{k=0}^{m} a_{m,k} \langle \mathbf{u}, x^k P_n \rangle = 0.$$

This proves (2.32). Next, for each $n \in \mathbb{N}$, multiplying both sides of (2.34) by x^{n-1}, we get

$$\langle \mathbf{u}, x^n P_n \rangle = \langle \mathbf{u}, x^{n-1} P_{n+1} \rangle + \beta_n \langle \mathbf{u}, x^{n-1} P_n \rangle + \gamma_n \langle \mathbf{u}, x^{n-1} P_{n-1} \rangle.$$

Hence, using (2.36),

$$\langle \mathbf{u}, x^n P_n \rangle = \gamma_n \langle \mathbf{u}, x^{n-1} P_{n-1} \rangle.$$

Applying successively this equality, we find

$$\langle \mathbf{u}, P_n^2 \rangle = \langle \mathbf{u}, x^n P_n \rangle = \gamma_0 \gamma_1 \cdots \gamma_n. \tag{2.37}$$

This holds for $n = 0$ since $\langle \mathbf{u}, 1 \rangle := \gamma_0$. Notice also that the first equality in (2.37) holds taking into account (2.36), after writing

$$P_n(x) = x^n + \sum_{k=0}^{n-1} a_{n,k} x^k.$$

It follows from (2.32) and (2.37) that $\mathbf{u}$ is regular and $\{P_n\}_{n \geq 0}$ is the corresponding monic OPS if and only if $\gamma_n \neq 0$ for each $n \in \mathbb{N}_0$. In addition, if $\mathbf{u}$ is positive definite and $\{P_n\}_{n \geq 0}$ is the corresponding monic OPS, then $\gamma_0 = \langle \mathbf{u}, 1 \rangle > 0$ and so, by Theorem 2.8, we may conclude that $\beta_n \in \mathbb{R}$ and $\gamma_n > 0$ for each $n \in \mathbb{N}_0$. Conversely, assume that $\beta_n \in \mathbb{R}$ and $\gamma_n > 0$ for each $n \in \mathbb{N}_0$. Then, by (2.30) we see that $P_n(x)$ is real (i.e., it has real coefficients) for each $n \in \mathbb{N}_0$. Moreover, from (2.37), we have $\langle \mathbf{u}, P_n^2 \rangle > 0$ for each $n \in \mathbb{N}_0$. This, together with (2.32), proves that $\{P_n\}_{n \geq 0}$ is the monic OPS with respect to $\mathbf{u}$. By Corollary 2.3, $\mathbf{u}$ is positive definite, and the theorem is proved. □

Since $\{P_n\}_{n \geq 0}$ in Theorem 2.9 is independent of γ_0, and $u_0 := \gamma_0$, the functional $\mathbf{u}$ is unique up to the (given) choice of γ_0, i.e., up to the choice of its first moment $u_0 := \langle \mathbf{u}, 1 \rangle$. The original Favard theorem concerned only the positive definite case and the functional $\mathbf{u}$ was represented by a Stieltjes integral. The corresponding result for regular functionals was subsequently observed by Shohat.

2.5 The Christoffel-Darboux Identity

In this section we state other important consequences of the three-term recurrence relation that characterizes a given OPS.

Theorem 2.10 (Christoffel-Darboux Identity). *Let $\{P_n\}_{n\geq 0}$ be a monic OPS fulfilling the three-term recurrence relation (2.20)–(2.21). Then, for each $n \in \mathbb{N}_0$,*

$$\sum_{j=0}^{n} \frac{P_j(x)P_j(y)}{\gamma_1\gamma_2\cdots\gamma_j} = \frac{1}{\gamma_1\gamma_2\cdots\gamma_n} \frac{P_{n+1}(x)P_n(y) - P_n(x)P_{n+1}(y)}{x-y} \quad \text{if} \quad x \neq y \tag{2.38}$$

(with the convention that the empty product equals one), and

$$\sum_{j=0}^{n} \frac{P_j^2(x)}{\gamma_1\gamma_2\cdots\gamma_j} = \frac{P'_{n+1}(x)P_n(x) - P'_n(x)P_{n+1}(x)}{\gamma_1\gamma_2\cdots\gamma_n}. \tag{2.39}$$

Proof. Since (2.39) follows from (2.38) by taking the limit $y \to x$, we only need to prove (2.38). From (2.20)–(2.21) we have

$$xP_n(x)P_n(y) = P_{n+1}(x)P_n(y) + \beta_n P_n(x)P_n(y) + \gamma_n P_{n-1}(x)P_n(y),$$

$$yP_n(y)P_n(x) = P_{n+1}(y)P_n(x) + \beta_n P_n(y)P_n(x) + \gamma_n P_{n-1}(y)P_n(x).$$

Subtracting the second equation from the first one yields

$$(x-y)P_n(x)P_n(y) = G_{n+1}(x,y) - \gamma_n G_n(x,y), \tag{2.40}$$

where

$$G_n(x,y) := P_n(x)P_{n-1}(y) - P_n(y)P_{n-1}(x).$$

Dividing both sides of (2.40) by $\gamma_1\gamma_2\cdots\gamma_n(x-y)$, and then in the resulting equality changing n into $j \in \mathbb{N}_0$, we obtain

$$\frac{P_j(x)P_j(y)}{\gamma_1\gamma_2\cdots\gamma_j} = \frac{G_{j+1}(x,y)}{\gamma_1\cdots\gamma_j(x-y)} - \frac{G_j(x,y)}{\gamma_1\cdots\gamma_{j-1}(x-y)}.$$

Summing from $j=0$ to $j=n$, the right-hand side becomes a telescoping sum, hence, taking into account that $G_0(x,y)=0$, we deduce (2.38), which completes the proof. □

Exercises

1. Is the simple set $\{x^n\}_{n\geq 0}$ an OPS with respect to some $\mathbf{u} \in \mathcal{P}'$?

2. Let $\mathbf{u} \in \mathcal{P}'$ be regular and let $\{P_n\}_{n\geq 0}$ be the corresponding monic OPS. Show that

$$P_n(x) = \frac{1}{\Delta_{n-1}} \begin{vmatrix} \langle \mathbf{u}, R_0R_0\rangle & \langle \mathbf{u}, R_0R_1\rangle & \cdots & \langle \mathbf{u}, R_0R_{n-1}\rangle & \langle \mathbf{u}, R_0R_n\rangle \\ \langle \mathbf{u}, R_1R_0\rangle & \langle \mathbf{u}, R_1R_1\rangle & \cdots & \langle \mathbf{u}, R_1R_{n-1}\rangle & \langle \mathbf{u}, R_1R_n\rangle \\ \vdots & \vdots & \ddots & \vdots & \vdots \\ \langle \mathbf{u}, R_{n-1}R_0\rangle & \langle \mathbf{u}, R_{n-1}R_1\rangle & \cdots & \langle \mathbf{u}, R_{n-1}R_{n-1}\rangle & \langle \mathbf{u}, R_{n-1}R_n\rangle \\ R_0(x) & R_1(x) & \cdots & R_{n-1}(x) & R_n(x) \end{vmatrix},$$

where $\{R_n\}_{n\geq 0}$ is any simple set of monic polynomials, and

$$\Delta_{-1} := 1, \quad \Delta_n := \det\left(\left[\langle \mathbf{u}, R_iR_j\rangle\right]_{i,j=0}^{n}\right).$$

3. Let $\{T_n\}_{n\geq 0}$ be the sequence of the Chebyshev polynomials of the first kind, defined by

$$T_n(x) = \cos(n\theta), \quad x = \cos\theta, \quad 0 \leq \theta \leq \pi, -1 \leq x \leq 1.$$

(a) Prove that $\{T_n\}_{n\geq 0}$ fulfills the three-term recurrence relation

$$2xT_n(x) = T_{n+1}(x) + T_{n-1}(x),$$

with initial conditions $T_0(x) = 1$ and $T_1(x) = x$.

(b) Set

$$p_0(x) := 1/\sqrt{\pi}\, T_0(x), \quad p_n(x) := \sqrt{2/\pi}\, T_n(x).$$

Show that $\{p_n\}_{n\geq 0}$ is orthonormal with respect to $\mathbf{u} \in \mathcal{P}'$ given by

$$\langle \mathbf{u}, p\rangle := \int_{-1}^{1} \frac{p(x)}{\sqrt{1-x^2}}\, dx.$$

(c) Prove that $T_n(x)$ admits the explicit expression

$$T_n(x) = \sum_{k=0}^{\lfloor n/2\rfloor} (-1)^k \binom{n}{2k} x^{n-2k}(1-x^2)^k,$$

where $\lfloor s\rfloor$ denotes the greatest integer less than or equal to the real number s.

4. Let $\{U_n\}_{n\geq 0}$ be the sequence of the Chebyshev polynomials of the second kind defined by

$$U_n(x) = \frac{\sin(n+1)\theta}{\sin\theta}, \quad x = \cos\theta, \quad 0 \leq \theta \leq \pi; -1 \leq x \leq 1.$$

(It is assumed that $U_n(x)$ is defined by continuity whenever $\sin\theta = 0$.)

   (a) Prove that $\{U_n\}_{n\geq 1}$ fulfills the three-term recurrence relation

   $$2xU_n(x) = U_{n+1}(x) + U_{n-1}(x),$$

   with initial conditions $U_0(x) = 1$ and $U_1(x) = 2x$.

   (b) Set $p_n(x) := \sqrt{2/\pi}\, U_n(x)$. Show that $\{p_n\}_{n\geq 0}$ is orthonormal with respect to $\mathbf{u} \in \mathcal{P}'$ given by

   $$\langle \mathbf{u}, p\rangle := \int_{-1}^{1} p(x)\sqrt{1-x^2}\mathrm{d}x.$$

   (c) Prove that $U_n(x)$ admits the explicit representation

   $$U_n(x) = \sum_{k=0}^{\lfloor (n+1)/2 \rfloor} (-1)^k \binom{n+1}{2k+1} x^{n-2k}(1-x^2)^k.$$

5. Let $\{P_n\}_{n\geq 0}$ be the sequence of Legendre polynomials defined by

$$P_n(x) := \frac{1}{2^n n!}\frac{\mathrm{d}^n}{\mathrm{d}x^n}\left\{(x^2-1)^n\right\}.$$

Notice that the leading coefficient of P_n is $2^{-n}\binom{2n}{n}$, hence it is not a monic polynomial. For each $n \in \mathbb{N}_0$, set

$$p_n(x) := \sqrt{\frac{2n+1}{2}}\, P_n(x).$$

Show that $\{p_n\}_{n\geq 0}$ is orthonormal with respect to $\mathbf{u} \in \mathcal{P}'$ given by

$$\langle \mathbf{u}, p\rangle := \int_{-1}^{1} p(x)\mathrm{d}x.$$

Chapter 3

Zeros and Gauss-Jacobi Mechanical Quadrature

In this chapter we present some basic properties of the zeros of an orthogonal polynomial sequence. In addition, we prove the Gauss-Jacobi mechanical quadrature formula and the Wendroff theorem. The methods of this chapter are quite elementary. In particular, no systematic use is made of the powerful tools from Matrix Analysis, from which relevant information concerning the zeros can also be derived. The main source for this chapter is the classical book by Chihara [14] and Ismail [23] and references therein.

3.1 Zeros of Orthogonal Polynomials

The zeros of an OPS when $\mathbf{u} \in \mathcal{P}'$ is positive definite exhibit a certain regularity in their behavior. In order to discuss this behavior we need to make an extension of the concept of positive definiteness as introduced in Definition 2.2. To make it clear we emphasize that a polynomial p is said to be nonzero on a set E (written $p \not\equiv 0$ on E) if it does not vanish identically on E.

Definition 3.1. *Let* $\mathbf{u} \in \mathcal{P}'$ *and* $E \subseteq \mathbb{R}$.

(i) $\mathbf{u}$ *is positive definite on* E *if*

$$\langle \mathbf{u}, p \rangle > 0, \tag{3.1}$$

for each real polynomial p *which is nonzero and nonnegative on* E*;*

(ii) *if* $\mathbf{u}$ *is positive definite on* E*, then* E *is called a supporting set for* $\mathbf{u}$.

DOI: 10.1201/9781003432067-3

Remark 3.1. *Notice that if $E = \mathbb{R}$ then positive definiteness on $\mathbb{R}$ is the same as positive definiteness as introduced in Definition 2.2.*

Theorem 3.1. *Let $E \subseteq \mathbb{R}$ with $\#E = \infty$. Let $\mathbf{u} \in \mathcal{P}'$ and suppose that $\mathbf{u}$ is positive definite on E. Then,*

(i) *if $E \subseteq S$, then $\mathbf{u}$ is positive definite on S;*

(ii) *if $E \supseteq S$ and $\overline{S} = E$, then $\mathbf{u}$ is positive definite on S.*

Proof. (i) Let p be a real polynomial and suppose that $p(x) \geq 0$, for all $x \in S$, and $p \not\equiv 0$ on S. Since $E \subseteq S$, $p(x) \geq 0$ for all $x \in E$ and, in addition, $p \not\equiv 0$ on E (since $p \not\equiv 0$ on S, p does not vanish identically on $\mathbb{R}$, and $\#E = \infty$). Therefore, since (by hypothesis) $\mathbf{u}$ is positive definite on E, we deduce $\langle \mathbf{u}, p \rangle > 0$, and so $\mathbf{u}$ is positive definite on S.

(ii) Take a real polynomial p such that $p(x) \geq 0$ for all $x \in S$ and $p \not\equiv 0$ on S. Then

$$p(x) \geq 0, \quad x \in E. \tag{3.2}$$

Indeed, suppose that there is $x_0 \in E$ with $p(x_0) < 0$. Since p is continuous on $\mathbb{R}$,

$$\exists\, \delta > 0 \;:\; x \in \mathbb{R},\; |x - x_0| < \delta \;\Rightarrow\; p(x) < 0. \tag{3.3}$$

Now, since $x_0 \in E$ and $\delta > 0$, taking into account that $\overline{S} = E$, we may ensure that

$$\exists\, s \in S \;:\; |s - x_0| < \delta. \tag{3.4}$$

From (3.3) and (3.4) we conclude that $p(s) < 0$, in contradiction with the choice of p. Henceforth, (3.2) holds. Moreover, $p \not\equiv 0$ on E, because $p \not\equiv 0$ on S and $S \subseteq E$. Thus, since (by hypothesis) $\mathbf{u}$ is positive definite on E, we conclude that $\langle \mathbf{u}, p \rangle > 0$. Thus, $\mathbf{u}$ is a positive definite on S. □

Remark 3.2. *Statement (ii) in Theorem 3.1 holds trivially if $\#E < \infty$, because in that case $S = E$. On the contrary, statement (i) does not hold if $\#E < \infty$. For instance, if $x_1, \ldots, x_N$ are any N distinct real numbers and $h_1, \ldots, h_N > 0$, then the functional $\mathbf{u} \in \mathcal{P}'$ defined by*

$$\langle \mathbf{u}, x^n \rangle := \sum_{j=1}^{N} h_j x_j^n, \quad n \in \mathbb{N}_0,$$

is positive definite on $E := \{x_1, \ldots, x_N\}$, but it is not positive definite *on any set $S \subseteq \mathbb{R}$ such that E is a proper subset of S. This can be seen immediately by noticing that $\mathbf{u}$ may be represented as a Riemann-Stieltjes integral with respect to the right continuous step function supported on E with jump h_j at the point x_j for each $j \in \{1, \dots, N\}$.*

Theorem 3.2. *Let $\mathbf{u} \in \mathcal{P}'$ be positive definite and let $\{P_n\}_{n\geq 0}$ be the monic OPS with respect to $\mathbf{u}$. Let I be an interval which is a supporting set for $\mathbf{u}$. Then, for each $n \in \mathbb{N}$, the zeros of P_n are all real, simple, and they are located in the interior of I.*

Proof. Fix $n \in \mathbb{N}$. Since $\mathbf{u}$ is positive definite, $P_n(x)$ is a real polynomial, i.e., its coefficients are real numbers (by Theorem 2.6). Moreover, since $\langle \mathbf{u}, P_n \rangle = 0$, $P_n(x)$ must change the sign at least once in the interior of the interval I. Indeed, if $P_n(x) \geq 0$ for all $x \in I$ then, since I is a supporting set for $\mathbf{u}$, we would have $\langle \mathbf{u}, P_n \rangle > 0$, a contradiction with $\langle \mathbf{u}, P_n \rangle = 0$. Thus, there is at least one point $r_1 \in I$ such that $P_n(r_1) < 0$. Similarly, if $P_n(x) \leq 0$ for all $x \in I$, then $-P_n(x) \geq 0$ for all $x \in I$, and we would have

$$\langle \mathbf{u}, P_n \rangle = -\langle \mathbf{u}, -P_n \rangle < 0,$$

again a contradiction with $\langle \mathbf{u}, P_n \rangle = 0$, and so there is at least one point $r_2 \in I$ such that $P_n(r_2) > 0$. Hence

$$P_n(r_1)P_n(r_2) < 0,$$

so $P_n(x)$ changes sign at least once in the interval $(r_1, r_2) \subset I$. Therefore, $P_n(x)$ has at least one zero of odd multiplicity located in the interior of I. Let $x_1, \dots, x_k$ denote the distinct zeros of an odd multiplicity of $P_n(x)$ which are located in the interior of I. Set

$$\pi_k(x) := (x - x_1)(x - x_2) \cdots (x - x_k).$$

Then the polynomial $\pi_k P_n$ has no zeros of odd multiplicity in the interior of I, hence $\pi_k(x)P_n(x) \geq 0$ for each $x \in I$. Therefore, since $\mathbf{u}$ is positive definite on I,

$$\langle \mathbf{u}, \pi_k P_n \rangle > 0. \tag{3.5}$$

On the other hand, since $\{P_n\}_{n\geq 0}$ is an OPS with respect to $\mathbf{u}$, we must have

$$\langle \mathbf{u}, \pi_k P_n \rangle \begin{cases} = 0 & \text{if} \quad k < n, \\ \neq 0 & \text{if} \quad k = n. \end{cases} \tag{3.6}$$

From (3.5) and (3.6) we deduce that $k = n$. This means that $P_n(x)$ has n

distinct zeros of odd multiplicity in the interior of I and, since $\deg P_n = n$, we may conclude that $P_n(x)$ has n real and simple zeros, all in the interior of I. □

Let $\{P_n\}_{n\geq 0}$ be the monic OPS with respect to a positive definite functional $\mathbf{u} \in \mathcal{P}'$. According to Theorem 3.2, the zeros $x_{n,1}, \ldots, x_{n,n}$ of each $P_n(x)$ may be ordered by increasing size, so that

$$x_{n,1} < x_{n,2} < \ldots < x_{n,n}. \tag{3.7}$$

Since $P_n(x)$ has positive leading coefficient $(= 1)$, it follows that

$$P_n(x) > 0 \qquad \text{if} \quad x > x_{n,n}; \tag{3.8}$$

$$\operatorname{sgn} P_n(x) = (-1)^n \qquad \text{if} \quad x < x_{n,1}, \tag{3.9}$$

where

$$\operatorname{sgn}(x) := \begin{cases} -1 & \text{if} \quad x < 0, \\ 0 & \text{if} \quad x = 0, \\ 1 & \text{if} \quad x > 0. \end{cases}$$

Theorem 3.3. *Let $\mathbf{u} \in \mathcal{P}'$ be positive definite and let $\{P_n\}_{n\geq 0}$ be the corresponding monic OPS. Suppose that the zeros of $P_n(x)$ fulfill (3.7) for each $n \geq 2$. Then the following holds:*

(i) *$P_n'(x)$ has exactly one zero in each open interval $(x_{n,j}, x_{n,j+1})$, $j = 1, 2, \ldots, n$. Moreover:*

$$\operatorname{sgn} P_n'(x_{n,j}) = (-1)^{n-j}; \tag{3.10}$$

(ii) *the zeros of $P_n(x)$ and $P_{n+1}(x)$ fulfill the interlacing property:*

$$x_{n+1,j} < x_{n,j} < x_{n+1,j+1}, \quad j = 1, 2, \ldots, n; \tag{3.11}$$

(iii) *for each $j \in \mathbb{N}$, $\{x_{n,j}\}_{n\geq j}$ is a decreasing sequence, while $\{x_{n,n-j+1}\}_{n\geq j}$ is an increasing sequence;*

(iv) *for each $j \in \mathbb{N}$, the limits*

$$\xi_j := \lim_{n\to\infty} x_{n,j}, \quad \eta_j := \lim_{n\to\infty} x_{n,n-j+1}, \tag{3.12}$$

all exist.

Proof. (i) Since $P_n(x)$ has n real and distinct zeros $x_{n,1}, \dots, x_{n,n}$, by the Cauchy-Bolzano theorem, $P_n'(x)$ has $n-1$ real and distinct zeros, one zero in between each pair of consecutive zeros of $P_n(x)$. Henceforth, in each interval $(x_{n,j}, x_{n,j+1})$ there is exactly one zero of $P_n'(x)$. Moreover, we see that for each $j \in \{1, 2, \dots, n-1\}$, $P_n'(x_{n,j})$ alternates in sign as j varies from 0 to n. Since $P_n'(x)$ has a positive leading coefficient, we conclude that (3.10) holds.

(ii) By the Cristoffel-Darboux formula (2.39), with $x = x_{n+1,j}$, we have

$$P'_{n+1}(x_{n+1,j}) P_n(x_{n+1,j}) > 0, \quad j = 1, 2, \dots, n+1. \tag{3.13}$$

On the other hand, by (3.10) with n replaced by $n+1$, we also have

$$\operatorname{sgn} P'_{n+1}(x_{n+1,j}) = (-1)^{n+1-j}, \quad j = 1, 2, \dots, n+1. \tag{3.14}$$

It follows from (3.13) and (3.14) that

$$\operatorname{sgn} P_n(x_{n+1,j}) = (-1)^{n+1-j}.$$

Therefore, $P_n(x)$ has at least one zero, and hence exactly one zero, on each of the intervals $(x_{n+1,j}, x_{n+1,j+1})$ which proves (3.11). The statement (iii) is an immediate consequence of (ii) and (iv) is an immediate consequence of (iii). □

Property (iv) in Theorem 3.3 motivates a very important definition:

Definition 3.2. *Let* $\mathbf{u} \in \mathcal{P}'$ *be positive definite and let* $\{P_n\}_{n\ge 0}$ *be the corresponding monic OPS. The closed interval* $[\xi, \eta]$*, where*

$$\xi := \lim_{n\to\infty} x_{n,1}, \quad \eta := \lim_{n\to\infty} x_{n,n}, \tag{3.15}$$

is called the true interval of orthogonality of $\{P_n\}_{n\ge 0}$.

Remark 3.3. *The true interval of orthogonality is the smallest closed interval that contains all the zeros of all the polynomials in the sequence* $\{P_n\}_{n\ge 0}$*. Moreover, it can be shown that the true interval of orthogonality is the smallest closed interval that is a supporting set for* $\mathbf{u}$.

Remark 3.4. *The three-term recurrence relation for a given monic OPS (not necessarily with respect to a positive definite moment linear functional)*

$$xP_{n-1}(x) = P_n(x) + \beta_{n-1}P_{n-1}(x) + \gamma_{n-1}P_{n-2}(x),$$

with initial conditions $P_{-1}(x) := 0$ *and* $P_0(x) = 1$, *may be written in matrix form as*

$$x\begin{bmatrix} P_0(x) \\ P_1(x) \\ \vdots \\ P_{n-2}(x) \\ P_{n-1}(x) \end{bmatrix} = J_n \begin{bmatrix} P_0(x) \\ P_1(x) \\ \vdots \\ P_{n-2}(x) \\ P_{n-1}(x) \end{bmatrix} + P_n(x)\begin{bmatrix} 0 \\ 0 \\ \vdots \\ 0 \\ 1 \end{bmatrix}, \tag{3.16}$$

where J_n *is a tridiagonal matrix of order* n *given by*

$$J_n := \begin{bmatrix} \beta_0 & 1 & & & & \\ \gamma_1 & \beta_1 & 1 & & & \\ & \gamma_2 & \beta_2 & 1 & & \\ & & \ddots & \ddots & \ddots & \\ & & & & \beta_{n-2} & 1 \\ & & & & \gamma_{n-1} & \beta_{n-1} \end{bmatrix}. \tag{3.17}$$

Clearly, for each $n \in \mathbb{N}$, *the following holds:*

(i) *The eigenvalues of* J_n *are the zeros of* P_n, *hence the spectrum of* J_n *is*

$$\sigma(J_n) = \{x_{n,j} : j = 1, \ldots, n\};$$

(ii) *an eigenvector* $v_{n,j}$ *corresponding to the eigenvalue* $x_{n,j}$ *is*

$$v_{n,j} := \begin{bmatrix} P_0(x_{n,j}) \\ P_1(x_{n,j}) \\ \vdots \\ P_{n-2}(x_{n,j}) \\ P_{n-1}(x_{n,j}) \end{bmatrix}, \quad j = 1, \ldots, n.$$

Remark 3.5. *Often we will refer to J_n as the Jacobi matrix associated with P_n, although in the framework of Linear Algebra the name "Jacobi" is usually attached to symmetric tridiagonal matrices.*

Remark 3.6. *As a consequence of the connection mentioned above, $P_n(x)$ is the (monic) characteristic polynomial associated with the matrix J_n, such that*

$$P_n(x) = \det\left(xI_n - J_n\right),$$

where I_n is the identity matrix of order n. Henceforth, $P_n(x)$ may be represented as a determinant involving only the sequences of the β and γ−parameters:

$$P_n(x) = \begin{vmatrix} x-\beta_0 & 1 & 0 & \cdots & 0 & 0 \\ \gamma_1 & x-\beta_1 & 1 & \cdots & 0 & 0 \\ 0 & \gamma_2 & x-\beta_2 & \cdots & 0 & 0 \\ \vdots & \vdots & \vdots & \ddots & \vdots & \vdots \\ 0 & 0 & 0 & \cdots & x-\beta_{n-2} & 1 \\ 0 & 0 & 0 & \cdots & \gamma_{n-1} & x-\beta_{n-1} \end{vmatrix}. \tag{3.18}$$

3.2 The Gauss-Jacobi Mechanical Quadrature

Fix n points $(t_j, y_j) \in \mathbb{R}^2$, $j = 1, \ldots, n$. Assume that $t_i \neq t_j$ if $i \neq j$. It is well known that the only solution for the problem—known as Lagrange problem—of constructing a polynomial of degree at most $n-1$ whose graph passes through all the points (t_j, y_j) is the so-called Lagrange interpolation polynomial, L_n, defined by

$$L_n(x) := \sum_{j=1}^{n} y_j \ell_j(x), \tag{3.19}$$

where

$$\ell_j(x) := \frac{F(x)}{(x-t_j)F'(t_j)}, \quad F(x) := \prod_{i=1}^{n}(x-t_i). \tag{3.20}$$

Clearly, ℓ_j is a polynomial of degree $n-1$ for each $j=1,\ldots,n$, which fulfills

$$\ell_j(t_k) = \delta_{j,k}, \quad k=1,2,\ldots,n. \tag{3.21}$$

Moreover, the interpolation property implies that L_n satisfies the property

$$L_n(t_j) = y_j, \quad j=1,2,\ldots,n. \tag{3.22}$$

We will use the Lagrange interpolation polynomial to obtain the Gauss-Jacobi mechanical quadrature formula.

Theorem 3.4 (Gauss-Jacobi Mechanical Quadrature). *Let* $\mathbf{u} \in \mathcal{P}'$ *be positive definite and let* $\{P_n\}_{n\geq 0}$ *be the corresponding monic OPS. For each* $n \in \mathbb{N}$*, denote by* $x_{n,1},\ldots,x_{n,n}$ *the zeros of* P_n*. Then*

$$\forall n \in \mathbb{N},\ \exists A_{n,1},\ldots,A_{n,n} > 0 \ :\ p \in \mathcal{P}_{2n-1},\ \langle \mathbf{u}, p\rangle = \sum_{j=1}^{n} A_{n,j} p(x_{n,j}). \tag{3.23}$$

Moreover,

$$\sum_{j=1}^{n} A_{n,j} = u_0 := \langle \mathbf{u}, 1\rangle. \tag{3.24}$$

Proof. Let $p \in \mathcal{P}_{2n-1}$. Consider the Lagrange interpolation polynomial L_n that passes through the points $(t_j, y_j) \equiv (x_{n,j}, p(x_{n,j}))$, $j=1,\ldots,n$, i.e.,

$$L_n(x) := \sum_{j=1}^{n} p(x_{n,j})\ell_{j,n}(x), \quad \ell_{j,n}(x) := \frac{P_n(x)}{(x-x_{n,j})P_n'(x_{n,j})}.$$

Let $Q(x) := p(x) - L_n(x)$. Thus $Q \in \mathcal{P}_{2n-1}$ and $Q(x_{n,j}) = p(x_{n,j}) - L_n(x_{n,j}) = y_j - y_j = 0$, and so $Q(x)$ vanishes at the zeros of $P_n(x)$. Therefore,

$$\exists R \in \mathcal{P}_{n-1} \ :\ Q(x) = R(x)P_n(x).$$

Since $\{P_n\}_{n\geq 0}$ is an OPS with respect to $\mathbf{u}$, we deduce

$$\langle \mathbf{u}, p\rangle = \langle \mathbf{u}, Q + L_n\rangle = \langle \mathbf{u}, RP_n\rangle + \langle \mathbf{u}, L_n\rangle = \langle \mathbf{u}, L_n\rangle = \sum_{j=1}^{n} p(x_{n,j})\,\langle \mathbf{u}, \ell_{j,n}\rangle.$$

Thus, setting

$$A_{n,j} := \langle \mathbf{u}, \ell_{j,n}\rangle, \tag{3.25}$$

we obtain

$$\langle \mathbf{u}, p \rangle = \sum_{j=1}^{n} A_{n,j} p(x_{n,j}). \tag{3.26}$$

Therefore, (3.23) will become proven provided we can show that the $A_{n,j}$'s defined by (3.25) are all positive numbers. Indeed, taking $p(x) \equiv \ell_{j,n}^2(x)$ in (3.26)—notice that each $\ell_{j,n}$ is a polynomial of degree $n-1$, hence $\ell_{j,n}^2 \in \mathcal{P}_{2n-1}$—, and taking into account that $\mathbf{u}$ is positive definite, we have

$$0 < \langle \mathbf{u}, \ell_{j,n}^2 \rangle = \sum_{k=1}^{n} A_{n,k} \ell_{j,n}^2(x_{n,k}) = \sum_{k=1}^{n} A_{n,k} \delta_{j,k} = A_{n,j}.$$

Notice also that the $A_{n,j}$'s defined by (3.25) do not depend on p. Hence, (3.23) is proved. Finally, choosing $p(x) \equiv 1$ in (3.23), we obtain (3.24), and the theorem is proved. □

Remark 3.7. *Quadrature formulas are very useful tools in Numerical Analysis, e.g. for computing integrals by approximation. Indeed, numerical quadrature consists of approximating the integral of a given integrable function f,*

$$I[f] := \int_{\mathbb{R}} f(x)\, \mathrm{d}\mu(x),$$

with respect to some positive Borel measure μ, by a finite sum which uses only the values of f at n points t_j (called nodes),

$$I_n[f] := \sum_{j=1}^{n} f(t_j) A_j,$$

where the coefficients A_j (which may depend on n, as well as the nodes t_j) have to be chosen properly so that the quadrature formula is correct,—i.e., the equality $I[f] = I_n[f]$ holds—, for as many functions f as possible.

3.3 The Wendroff Theorem

The following theorem was stated as a footnote by Geronimus [20] and it was proved by Wendroff [53].

Theorem 3.5 (Wendroff). *Given real numbers* $x_1 < x_2 < \cdots < x_{n+1}$ *and* $y_1 < y_2 < \cdots < y_n$ *such that*

$$x_j < y_j < x_{j+1}, \quad j = 1, \ldots, n,$$

there exists an OPS, $\{P_n\}_{n\geq 0}$, *with respect to a positive definite linear functional, such that*

$$P_n(x) = (x - y_1)(x - y_2) \cdots (x - y_n), \tag{3.27}$$

$$P_{n+1}(x) = (x - x_1)(x - x_2) \cdots (x - x_{n+1}). \tag{3.28}$$

Proof. We shall construct a set of polynomials, $\{P_n\}_{n\geq 0}$, with $P_0 = 1$ and P_n and P_{n+1} given by (3.27) and (3.28) respectively, such that (2.20) and (2.23) hold. The real number β_n is defined by the requirement that the polynomial

$$R_{n-1}(x) := (x - \beta_n)P_n(x) - P_{n+1}(x) \in \mathcal{P}_{n-1}.$$

Note that

$$\operatorname{sgn} R_{n-1}(y_j) = -\operatorname{sgn} P_{n+1}(y_j), \quad j = 1, \ldots, n,$$

and so, by (3.11), R_{n-1} has at least n sign changes, so it must have degree $n-1$ and its zeros interlace with the zeros of P_n. Now, the real number γ_n is defined by the requirement that the polynomial $P_{n-1} := R_{n-1}/\gamma_n$ be monic. Finally, by (3.11), we get

$$\gamma_n = -\frac{P_{n+1}(y_n)}{P_{n-1}(y_n)} > 0.$$

We can now repeat this procedure to obtain $P_{n-2}, P_{n-3}, \ldots, P_1, P_0 = 1$. To obtain the remaining elements of $\{P_n\}_{n\geq 0}$ we use (2.20) and (2.23), and the result follows from the Favard theorem. □

Remark 3.8. *From Remark 3.4, we see that the Wendroff theorem can be reformulated in terms of embedding conditions for real symmetric tridiagonal matrices. Certainly, Wendroff was not aware that previous results had already been published, because before publishing his paper on the subject, Fan and Pall had already proved the following theorem [16]: Let* A *and* B *be two Hermitian matrices of respective orders* n *and* m, *where* $n \geq m$. *Let* $\alpha_1 \leq \alpha_2 \leq \cdots \leq \alpha_n$ *and* $\beta_1 \leq \beta_2 \leq \cdots \leq \beta_m$ *be the eigenvalues of* A *and* B, *respectively. Then a necessary and sufficient condition for* B *to be embeddable in* A *is that inequalities*

$$\alpha_j \leq \beta_j, \quad \alpha_{n-j+1} \geq \beta_{m-j+1}, \quad j = 1, \ldots, m,$$

be fulfilled. (Recall that B is embeddable in A if there exists a unitary matrix U of order n such that U^*AU contains B as a principal submatrix. In the case of real symmetric matrices A and B, this theorem remains valid if we require U to be orthogonal in the definition of embedding.)

The following result was proved by Chamberland [12]. Here we present a simple proof is given in [9], which makes use of the Wendroff theorem.

Theorem 3.6. *Let P be a polynomial of degree $n \geq 2$ with real coefficients. Then the zeros of P are real and simple if and only if*

$$\left(\left(P^{(n-j-1)}(x)\right)'\right)^2 - P^{(n-j-1)}(x)\left(P^{(n-j-1)}(x)\right)'' > 0, \quad j = 1, \ldots, n-1, \tag{3.29}$$

for all real x. (Here $P^{(k)}$ is the derivative of order $k \in \mathbb{N}$ of P.)

Proof. Assume that the zeros of P are real and simple. By the Wendroff theorem, there are positive definite linear functionals $\mathbf{u}_j$ so that $P_{j+1} = P^{(n-j-1)}$ and $P_j = P'_{j+1}$ are among the orthogonal polynomials for $\mathbf{u}_j$. Since P_{j+1} and P_j have leading coefficients of the same sign, by the Christoffel-Darboux identity (2.39), we have

$$P'_{j+1}(x)P_j(x) - P'_j(x)P_{j+1}(x) > 0, \quad 1 \leq j \leq n-1, \tag{3.30}$$

for all real x. Of course, (3.30) makes it obvious that (3.29) holds. Suppose that (3.29) holds. Set $P_j = P^{(n-j)}$ for all $j = 1, \ldots, n-1$. Without loss of generality, we can assume that P is monic. Suppose that the zeros of P_j are real and simple. From (3.30) we see that P_{j+1} has at least one zero between two zeros of P_j, and so P_{j+1} has at least $j-1$ real and simple zeros. Suppose that the other two zeros of P_{j+1} are not real, and therefore they appear as a complex conjugate pair. Let a be the largest zero of P_j. Clearly, $P_{j+1}(x) > 0$ for all $x > a$. However, by (3.30), we have $P_{j+1}(a) < 0$, which leads to a contradiction. From this we conclude that if P_j has real and simple zeros, then the same holds for P_{j+1}. Since $P_1 = P^{(n-1)}$ has a single real zero, we infer successively that $P_2, \ldots, P_n = P$ have real and simple zeros, and the result follows.

□

Exercises

1. Let $x_1, \ldots, x_N$ be any N distinct real numbers, and let $h_1, \ldots, h_N > 0$. Define $\mathbf{u} \in \mathcal{P}'$ by

$$\langle \mathbf{u}, x^n \rangle := \sum_{j=1}^{N} h_j x_j^n, \quad n \in \mathbb{N}_0.$$

Prove that:

(a) $\mathbf{u}$ is positive definite on $E := \{x_1, \ldots, x_N\}$;

(b) $\mathbf{u}$ is not positive definite on any set $S \subseteq \mathbb{R}$ such that E is a proper subset of S.

2. Let $a, b, c \in \mathbb{R}$, with $bc > 0$. For each $n \in \mathbb{N}_0$, set

$$P_n(x) := (bc)^{n/2} U_n\left(\frac{x-a}{2\sqrt{bc}}\right),$$

where $\{U_n\}_{n\geq 0}$ is the sequence of the Chebyshev polynomials of the second kind.

(a) Show that $\{P_n\}_{n\geq 0}$ is a monic OPS with respect to a positive definite functional $\mathbf{u} \in \mathcal{P}'$.

(b) Consider the tridiagonal Toeplitz matrix of order n

$$A_n = \begin{bmatrix} a & b & & & & \\ c & a & b & & & \\ & c & a & b & & \\ & & \ddots & \ddots & \ddots & \\ & & & c & a & b \\ & & & & c & a \end{bmatrix}.$$

Prove that the eigenvalues of A_n are

$$\lambda_j := a + 2\sqrt{bc}\cos\frac{j\pi}{n+1}, \quad j = 1, 2, \ldots, n,$$

with corresponding eigenvectors

$$v_j := \frac{1}{\sin\dfrac{j\pi}{n+1}} \begin{bmatrix} \sin\dfrac{j\pi}{n+1} \\ (c/b)^{1/2}\sin\dfrac{2j\pi}{n+1} \\ \vdots \\ (c/b)^{(n-1)/2}\sin\dfrac{nj\pi}{n+1} \end{bmatrix}.$$

3. Let $\mathbf{u} \in \mathcal{P}'$ be positive definite. Let $\{P_n\}_{n\geq 0}$ be the corresponding monic OPS and let $\{p_n\}_{n\geq 0}$ be the associated orthonormal sequence. Denote by $x_{n,1}, \ldots, x_{n,n}$ the zeros of P_n and let $\{\gamma_n\}_{n\geq 1}$ be the sequence of $\gamma-$parameters appearing in the three-term recurrence relation fulfilled by $\{P_n\}_{n\geq 0}$. Show that the "weights" $A_{n,k}$ in the associated Gauss-Jacobi mechanical quadrature formula admit the following representations:

$$A_{nk} = -\frac{u_0\gamma_1\gamma_2\cdots\gamma_n}{P_{n+1}(x_{n,k})P_n'(x_{n,k})} = \left(\sum_{j=0}^{n} p_j^2(x_{nk})\right)^{-1}, \quad k = 1, \ldots, n.$$

4. Let $\mathbf{u} \in \mathcal{P}'$ be defined as

$$\langle \mathbf{u}, p\rangle := \int p(x)\, \mathrm{d}\mu(x), \quad p \in \mathcal{P},$$

where μ is a positive Borel measure with infinite support and finite moments of all orders. (Recall that the support of μ is the set $\mathrm{supp}(\mu) := \{x \in \mathbb{R} : \mu((x-\epsilon, x+\epsilon)) > 0,\ \epsilon > 0\}$.)

   (a) Prove that $\mathbf{u}$ is positive definite.

   (b) Let $\{P_n\}_{n\geq 0}$ be the monic OPS with respect to $\mathbf{u}$. Prove that the maximum of the ratio

$$\frac{\displaystyle\int xQ_n^2(x)\, \mathrm{d}\mu(x)}{\displaystyle\int Q_n^2(x)\, \mathrm{d}\mu(x)}$$

   taken over all real polynomials Q_n of degree at most n is equal to the largest zero $x_{n+1,n+1}$ of the polynomial P_{n+1}, and the minimum is equal to the smallest zero $x_{n+1,1}$ of P_{n+1}.

   (c) Determine polynomials Q_n where these maximum and minimum ratios are attained.

Chapter 4

The Spectral Theorem

In this chapter we concentrate on the study of orthogonal polynomials sequences with respect to positive definite moment linear functionals. The main objective is to prove that these functionals admit an integral representation involving a positive Borel measure on the real line.

4.1 The Helly Theorems

In this section we state some preliminary results needed for the proof of the representation theorem to be stated in the next section.

Lemma 4.1. *Let $I \subseteq \mathbb{R}$ be an interval and let $f : I \to \mathbb{R}$ be a monotone function. Then f has at most countably many discontinuity points.*

The above lemma is a well-known result. A concise proof of this result appears in [29].

Lemma 4.2. *Let $\{f_n\}_{n\geq 1}$ be a sequence of real functions defined on a countable set E. Suppose that $\{f_n(x)\}_{n\geq 1}$ is a bounded sequence for each $x \in E$. Then $\{f_n\}_{n\geq 1}$ contains a subsequence $\{f_{n_j}\}_{j\geq 1}$ that converges everywhere on E, i.e., the (sub)sequence $\{f_{n_j}(x)\}_{j\geq 1}$ converges for each $x \in E$.*

Proof. Set $E := \{x_1, x_2, x_3, \ldots\}$ and set $f_n^{(0)} \equiv f_n$. Since $\{f_n^{(0)}(x_1)\}_{n\geq 1}$ is a bounded sequence of real numbers, it contains a convergent subsequence, i.e., there exists a subsequence $\{f_n^{(1)}\}_{n\geq 1}$ of $\{f_n^{(0)}\}_{n\geq 1}$ such that $\{f_n^{(1)}(x)\}_{n\geq 1}$ converges for $x = x_1$. Now, since $\{f_n^{(1)}(x_2)\}_{n\geq 1}$ is a bounded sequence, it contains a convergent subsequence. Theorefore, there exists a subsequence $\{f_n^{(2)}\}_{n\geq 1}$

DOI: 10.1201/9781003432067-4

of $\{f_n^{(1)}\}_{n\geq 1}$ such that $\{f_n^{(2)}(x)\}_{n\geq 1}$ converges for $x = x_2$. Proceeding in this way, we obtain sequences

$$\{f_n^{(0)}\}_{n\geq 1}, \{f_n^{(1)}\}_{n\geq 1}, \{f_n^{(2)}\}_{n\geq 1}, \ldots, \{f_n^{(k)}\}_{n\geq 1}, \ldots$$

such that:

(i) $\{f_n^{(k)}\}_{n\geq 1}$ is a subsequence of $\{f_n^{(k-1)}\}_{n\geq 1}$ for each $k = 1, 2, 3, \ldots$;

(ii) $\{f_n^{(k)}(x)\}_{n\geq 1}$ converges for each $x \in E_k := \{x_1, x_2, \ldots, x_k\}$.

It follows from (i) —with a little care (passing to a subsequence if necessary) to ensure that the relative order of terms is preserved—that the diagonal sequence, $\{f_n^{(n)}\}_{n\geq 1}$, is also a subsequence of $\{f_n\}_{n\geq 1}$. Since, for each $k \in \mathbb{N}$, except for the first $k-1$ terms, $\{f_n^{(n)}\}_{n\geq 1}$ is also a subsequence of $\{f_n^{(k)}\}_{n\geq 1}$, it follows from (ii) that

$$\{f_n^{(n)}(x)\}_{n\geq 1} \text{ converges for each } x \in \bigcup_{k=1}^{\infty} E_k = E.$$

Finally, since $\{f_n^{(n)}(x)\}_{n\geq 1}$ is a subsequence of $\{f_n(x)\}_{n\geq 1}$ for each $x \in E$, the result follows. □

Theorem 4.1 (Helly Selection Principle). *Let $\{\phi_n\}_{n\geq 1}$ be a uniformly bounded sequence of nondecreasing functions defined on $\mathbb{R}$. Then, $\{\phi_n\}_{n\geq 1}$ has a subsequence which converges on $\mathbb{R}$ to a bounded and nondecreasing function.*

Proof. Consider the set of rational numbers, $\mathbb{Q}$. According to Lemma 4.2, there is a subsequence $\{\phi_{n_k}\}_{k\geq 1}$ which converges everywhere on $\mathbb{Q}$. Henceforth, we may define a function $\Phi_1 : \mathbb{Q} \to \mathbb{R}$ as

$$\Phi_1(r) := \lim_{k\to+\infty} \phi_{n_k}(r), \quad r \in \mathbb{Q}. \tag{4.1}$$

It follows from the hypothesis on $\{\phi_n\}_{n\geq 1}$ that Φ_1 is bounded and nondecreasing on $\mathbb{Q}$. We now extend the domain of Φ_1 to $\mathbb{R}$ by defining $\Phi_2 : \mathbb{R} \to \mathbb{R}$ as

$$\Phi_2(x) := \begin{cases} \Phi_1(x), & x \in \mathbb{Q}, \\ \sup\limits_{\substack{r\in\mathbb{Q} \\ r<x}} \Phi_1(r), & x \in \mathbb{R}\setminus\mathbb{Q}. \end{cases} \tag{4.2}$$

The function Φ_2 is clearly bounded and nondecreasing on $\mathbb{R}$. According to (4.1), $\{\phi_{n_k}\}_{k\geq 1}$ converges to $\Phi_2(x)$ at each point $x \in \mathbb{Q}$. We next claim that

$\{\phi_{n_k}\}_{k\geq 1}$ also converges to $\Phi_2(x)$ at each point x where Φ_2 is continuous. Indeed, suppose that Φ_2 is continuous at the point $x \in \mathbb{R}\setminus\mathbb{Q}$. Since $\mathbb{Q}$ is a dense subset of $\mathbb{R}$ and Φ_2 is continuous at x, then

$$\forall\, \epsilon > 0,\ \exists\, x_2 \in \mathbb{Q} \,:\, x < x_2 \ \wedge \ \Phi_2(x_2) < \Phi_2(x) + \epsilon. \tag{4.3}$$

Fix arbitrarily $x_1 \in \mathbb{Q}$, with $x_1 < x$. Then, since (by hypothesis) ϕ_{n_k} is a nondecreasing function on $\mathbb{R}$, we have

$$\phi_{n_k}(x_1) \leq \phi_{n_k}(x) \leq \phi_{n_k}(x_2).$$

Therefore, we deduce

$$\begin{aligned}
\Phi_2(x_1) = \Phi_1(x_1) &= \lim_{k\to+\infty} \phi_{n_k}(x_1) \\
&= \liminf_{k\to+\infty} \phi_{n_k}(x_1) \leq \liminf_{k\to+\infty} \phi_{n_k}(x) \\
&\leq \limsup_{k\to+\infty} \phi_{n_k}(x) \leq \limsup_{k\to+\infty} \phi_{n_k}(x_2) = \lim_{k\to+\infty} \phi_{n_k}(x_2) \\
&= \Phi_1(x_2) = \Phi_2(x_2) \\
&< \Phi_2(x) + \epsilon.
\end{aligned}$$

Summarizing, we proved that, if Φ_2 is continuous at a point $x \in \mathbb{R}\setminus\mathbb{Q}$, then

$$x_1 \in \mathbb{Q}, \quad x_1 < x \ \Rightarrow\ \Phi_2(x_1) \leq \liminf_{k\to+\infty} \phi_{n_k}(x) \leq \limsup_{k\to+\infty} \phi_{n_k}(x) < \Phi_2(x) + \epsilon.$$

Therefore,

$$\begin{aligned}
\Phi_2(x) = \lim_{\substack{x_1\to x^- \\ x_1\in\mathbb{Q}}} \Phi_2(x_1) &\leq \liminf_{k\to+\infty} \phi_{n_k}(x) \\
&\leq \limsup_{k\to+\infty} \phi_{n_k}(x) < \Phi_2(x) + \epsilon,
\end{aligned}$$

hence, since $\epsilon > 0$ is arbitrary, we deduce (letting $\epsilon \to 0^+$),

$$\Phi_2(x) \leq \liminf_{k\to+\infty} \phi_{n_k}(x) \leq \limsup_{k\to+\infty} \phi_{n_k}(x) \leq \Phi_2(x).$$

Since the left-hand side and the right-hand side coincide, these inequalities are indeed equalities, hence

$$\lim_{k\to+\infty} \phi_{n_k}(x) = \Phi_2(x).$$

Thus $\{\phi_{n_k}(x)\}_{k\geq 1}$ converges to $\Phi_2(x)$ at each point $x \in \mathbb{R}$ of continuity of Φ_2. Now, observe that Φ_2 is a nondecreasing function, so (by Lemma 4.1) the set of its points of discontinuity form an at most countable set. Denote by D the set of points of discontinuity of Φ_2 which does not belong to $\mathbb{Q}$. Applying Lemma 4.2 to $\{\phi_{n_k}\}_{k\geq 1}$ and D, we deduce that there is a subsequence $\{\phi_{n_{k_j}}\}_{j\geq 1}$ of $\{\phi_{n_k}\}_{k\geq 1}$ which converges everywhere on D to a limit function $\Phi_3 : D \to \mathbb{R}$. Finally, define $\phi : \mathbb{R} \to \mathbb{R}$ by

$$\phi(x) := \begin{cases} \Phi_2(x), & x \in \mathbb{R} \setminus D, \\ \Phi_3(x), & x \in D. \end{cases}$$

It is clear that

$$\lim_{j\to+\infty} \phi_{n_{k_j}}(x) = \phi(x), \quad x \in \mathbb{R}.$$

Moreover, the hypothesis on $\{\phi_n\}_{n\geq 1}$ ensures that ϕ is bounded and nondecreasing. In conclusion: $\{\phi_n\}_{n\geq 1}$ has a convergent subsequence $\{\phi_{n_{k_j}}\}_{j\geq 1}$ which converges to a bounded and nondecreasing function ϕ on $\mathbb{R}$, and the theorem follows. □

The next theorem involves the Riemann-Stieltjes integral. The needed facts concerning this integral can be found in [26].

Theorem 4.2 (Helly Convergence Theorem). *Let $\{\phi_n\}_{n\geq 1}$ be a uniformly bounded sequence of nondecreasing functions defined on a compact interval $[a, b]$. Suppose that $\{\phi_n\}_{n\geq 1}$ converges on $[a, b]$ to a limit function ϕ, so that*

$$\phi(x) := \lim_{n\to\infty} \phi_n(x), \quad x \in [a, b]. \tag{4.4}$$

Then, for each continuous function $f : [a, b] \to \mathbb{R}$, the following holds:

$$\lim_{n\to\infty} \int_a^b f(x)\, \mathrm{d}\phi_n(x) = \int_a^b f(x)\, \mathrm{d}\phi(x). \tag{4.5}$$

Proof. Since $\{\phi_n\}_{n\geq 1}$ is uniformly bounded and each ϕ_n is nondecreasing on $[a, b]$,

$$\exists M > 0 \;:\; \forall n \in \mathbb{N}, \quad 0 \leq \phi_n(b) - \phi_n(a) \leq M.$$

Hence, by hypothesis (4.4), we also have

$$0 \leq \phi(b) - \phi(a) \leq M. \tag{4.6}$$

Fix $\epsilon > 0$. Since, by hypothesis, f is a continuous function on the compact set $[a, b]$, then f is uniformly continuous on $[a, b]$, and so there is a partition P_ϵ of $[a, b]$,

$$P_\epsilon \;:\; a = x_0 < x_1 < \cdots < x_{j-1} < x_j < \cdots < x_{m-1} < x_m = b,$$

such that

$$x, z \in [x_{j-1}, x_j] \quad \Rightarrow \quad \left|f(x) - f(z)\right| < \frac{\epsilon}{2M}. \tag{4.7}$$

Indeed, f being uniformly continuous on $[a, b]$, this means that

$$\forall \epsilon > 0,\ \exists\, \delta = \delta(\epsilon) > 0 \ :\ x, z \in [a, b], \quad |x - z| < \delta \ \Rightarrow\ \left|f(x) - f(z)\right| < \frac{\epsilon}{2M};$$

hence we choose the partition P_ϵ so that $|x_j - x_{j-1}| < \delta$. Now, choose an "intermediate point" $\xi_j \in [x_{j-1}, x_j]$ and set

$$\Delta_j \phi := \phi(x_j) - \phi(x_{j-1}), \quad \Delta_j \phi_n := \phi_n(x_j) - \phi_n(x_{j-1}).$$

By the Mean Value Theorem for the Riemann-Stieltjes integral, there exists $\xi_j' \in [x_{j-1}, x_j]$ such that

$$\begin{aligned} \int_{x_{j-1}}^{x_j} f(x)\, \mathrm{d}\phi(x) &= f(\xi_j')\left(\phi(x_j) - \phi(x_{j-1})\right) \\ &= \left(f(\xi_j') - f(\xi_j) + f(\xi_j)\right) \Delta_j \phi. \end{aligned}$$

Hence

$$\int_{x_{j-1}}^{x_j} f(x)\, \mathrm{d}\phi(x) - f(\xi_j)\Delta_j\phi = \left(f(\xi_j') - f(\xi_j)\right)\Delta_j\phi.$$

Summing over j and taking into account that

$$\sum_{j=1}^{m} \int_{x_{j-1}}^{x_j} f(x)\, \mathrm{d}\phi(x) = \int_a^b f(x)\, \mathrm{d}\phi(x),$$

and then applying the triangular inequality, we get

$$\left| \int_a^b f(x)\, \mathrm{d}\phi(x) - \sum_{j=1}^{m} f(\xi_j)\Delta_j\phi \right| \le \sum_{j=1}^{m} \left|f(\xi_j') - f(\xi_j)\right|\Delta_j\phi.$$

Since $\xi_j, \xi_j' \in [x_{j-1}, x_j]$, so that we can apply (4.7), we obtain

$$\begin{aligned} \left| \int_a^b f(x)\, \mathrm{d}\phi(x) - \sum_{i=1}^{m} f(\xi_j)\Delta_j\phi \right| &< \frac{\epsilon}{2M} \sum_{j=1}^{m} \Delta_j\phi \\ &= \frac{\epsilon}{2M}\left(\phi(b) - \phi(a)\right) \le \frac{\epsilon}{2}, \end{aligned} \tag{4.8}$$

where the last inequality follows from (4.6). In the same way, replacing ϕ by ϕ_n in the previous reasoning, we deduce

$$\left| \int_a^b f(x)\, \mathrm{d}\phi_n(x) - \sum_{j=1}^{m} f(\xi_j)\Delta_j\phi_n \right| < \frac{\epsilon}{2}. \tag{4.9}$$

Next, observe that

$$\left| \int_a^b f(x)\,\mathrm{d}\phi(x) - \int_a^b f(x)\,\mathrm{d}\phi_n(x) \right|$$

$$\leq \left| \int_a^b f(x)\,\mathrm{d}\phi(x) - \sum_{j=1}^{m} f(\xi_j)\Delta_j\phi \right| + \left| \sum_{j=1}^{m} f(\xi_j)\Delta_j\phi - \sum_{j=1}^{m} f(\xi_j)\Delta_j\phi_n \right|$$

$$+ \left| \int_a^b f(x)\,\mathrm{d}\phi_n(x) - \sum_{j=1}^{m} f(\xi_j)\Delta_j\phi_n \right| .$$

Therefore, taking into account (4.8) and (4.9), and noticing that

$$\left| \sum_{j=1}^{m} f(\xi_j)\Delta_j\phi - \sum_{j=1}^{m} f(\xi_j)\Delta_j\phi_n \right| \leq \sum_{i=1}^{m} |f(\xi_j)|\ |\Delta_j\phi - \Delta_j\phi_n| ,$$

we obtain

$$\left| \int_a^b f(x)\,\mathrm{d}\phi(x) - \int_a^b f(x)\,\mathrm{d}\phi_n(x) \right| < \epsilon + \sum_{i=1}^{m} |f(\xi_j)|\ |\Delta_j\phi - \Delta_j\phi_n| . \quad (4.10)$$

Keeping P_ϵ fixed, we have

$$\lim_{n\to+\infty} \left(\Delta_j\phi - \Delta_j\phi_n\right) = \lim_{n\to+\infty} \left\{ \left(\phi(x_j) - \phi_n(x_j)\right) - \left(\phi(x_{i-1}) - \phi_n(x_{i-1})\right) \right\} = 0,$$

where the last equality follows from (4.4). Therefore, from (4.10) we get

$$\limsup_{n\to+\infty} \left| \int_a^b f(x)\,\mathrm{d}\phi(x) - \int_a^b f(x)\,\mathrm{d}\phi_n(x) \right| \leq \epsilon. \quad (4.11)$$

Since ϵ is positive and arbitrary, the limit superior in (4.11) must be equal to zero. It turns out that in (4.11) we may replace the limit superior by the limit, and since this limit is equal to zero, we conclude that (4.5) holds, and the theorem is proved. □

Lemma 4.2 and the Helly theorems can be found in Chihara's book [14].

4.2 The Representation Theorem

We are ready to state the important representation theorem for a positive definite functional, showing that such a functional admits an integral representation as a Riemann-Stieltjes integral with respect to a real-bounded nondecreasing function on $\mathbb{R}$ fulfilling some natural conditions (namely, finite moments of all orders and infinite spectrum). We begin by introducing some useful concepts.

Definition 4.1. *A function $\psi : \mathbb{R} \to \mathbb{R}$ is called a distribution function if it is bounded, nondecreasing and all its moments*

$$\int_{-\infty}^{+\infty} x^n \, \mathrm{d}\psi(x), \quad n \in \mathbb{N}_0,$$

are finite. The spectrum of a distribution function ψ is the set

$$\sigma(\psi) := \big\{ x \in \mathbb{R} \, : \, \psi(x+\delta) - \psi(x-\delta) > 0, \, \delta > 0 \big\}.$$

Remark 4.1. *Often, ψ being a distribution function, a point in $\sigma(\psi)$ is called a spectral point, or an increasing point of ψ.*

Theorem 4.3. *The spectrum $\sigma(\psi)$ of a distribution function ψ is closed in $\mathbb{R}$.*

Proof. We will prove that $\mathbb{R} \setminus \sigma(\psi)$ is an open set. Let $x_0 \in \mathbb{R} \setminus \sigma(\psi)$. Then

$$\exists \, \delta > 0 \, : \, \psi(x_0+\delta) - \psi(x_0-\delta) \leq 0.$$

Since ψ is nondecreasing, ψ must be constant ($= C$) on the interval $(x_0 - \delta, x_0 + \delta)$. Therefore, we see that if $x \in \big(x_0 - \delta/2, x_0 + \delta/2\big)$, then there is $\delta' > 0$ (choose δ' such that $0 < \delta' < \delta/2$) such that

$$\psi(x+\delta') - \psi(x-\delta') = C - C = 0,$$

and so $x \notin \sigma(\psi)$. Thus,

$$\left(x_0 - \frac{\delta}{2}, x_0 + \frac{\delta}{2} \right) \subseteq \mathbb{R} \setminus \sigma(\psi),$$

so that x_0 is an interior point of $\mathbb{R} \setminus \sigma(\psi)$. Since x_0 was arbitrarily fixed on the set $\mathbb{R} \setminus \sigma(\psi)$, we conclude that this set is open in $\mathbb{R}$, and the theorem follows. □

Let $\mathbf{u} \in \mathcal{P}'$ be positive definite and $\{P_n\}_{n\geq 0}$ the monic OPS with respect to $\mathbf{u}$. By Theorems 2.6 and 3.2, each $P_n(x)$ is a real polynomial having n real simple zeros:

$$x_{n,1} < x_{n,2} < \cdots < x_{n,n}.$$

For each $n \in \mathbb{N}$, let us introduce a distribution function $\psi_n : \mathbb{R} \to \mathbb{R}$, characterized as being a bounded and right continuous step function with a spectrum $\sigma(\psi_n) = \{x_{n,1}, \ldots, x_{n,n}\}$, and having jump $A_{n,j} > 0$ at the jth zero $x_{n,j}$ of $P_n(x)$, where the $A_{n,j}$'s are the weights appearing in Gauss-Jacobi mechanical quadrature formula, so that

$$\begin{aligned} &\psi_n(-\infty) := 0, \quad \psi_n(+\infty) := u_0, \\ &\psi_n(x_{n,j}) - \psi_n(x_{n,j} - 0) = A_{n,j}, \quad j = 1, \ldots, n. \end{aligned} \tag{4.12}$$

Explicitly, we may write

$$\psi_n(x) := \begin{cases} 0, & x < x_{n,1}; \\ A_{n,1} + \cdots + A_{n,j}, & x_{n,j} \leq x < x_{n,j+1}, \quad j = 1, \ldots, n-1; \\ u_0, & x \geq x_{n,n}. \end{cases} \tag{4.13}$$

Then, for each fixed $k \in \{0, 1, \ldots, 2n-1\}$, by using the Gauss-Jacobi-Christoffel quadrature formula (Theorem 3.4) applied to the polynomial $p(x) := x^k$, one sees that the moment $u_k := \langle \mathbf{u}, x^k \rangle$ may be represented as a Riemann-Stieltjes integral with respect to ψ_n as:

$$u_k = \sum_{j=1}^{n} A_{n,j} x_{n,j}^k = \int_{-\infty}^{+\infty} x^k \, d\psi_n(x), \quad k = 0, 1, \ldots, 2n-1. \tag{4.14}$$

Now, by the Helly selection principle, there is a subsequence $\{\psi_{n_j}\}_{j\geq 0}$ of $\{\psi_n\}_{n\geq 0}$ which converges on $\mathbb{R}$ to a bounded and nondecreasing function ψ:

$$\psi(x) := \lim_{j\to\infty} \psi_{n_j}(x), \quad x \in \mathbb{R}. \tag{4.15}$$

Definition 4.2. *Let $\mathbf{u} \in \mathcal{P}'$ be positive definite. A function $\psi : \mathbb{R} \to \mathbb{R}$ defined as in (4.15)—limit of a subsequence of the step functions (4.12)—is called a natural representative for $\mathbf{u}$.*

Remark 4.2. *A natural representative ψ for $\mathbf{u}$ is a distribution function.*

Indeed, as noted above, ψ is bounded and nondecreasing. Moreover, all the moments

$$\int_{-\infty}^{+\infty} x^n \, d\psi(x), \quad n \in \mathbb{N}_0,$$

are finite, as follows by the representation Theorem 4.4, to be proved next.

Theorem 4.4 (Representation Theorem). *Let $\mathbf{u} \in \mathcal{P}'$ be positive definite. Then, there is a natural representative of $\mathbf{u}$, $\psi : \mathbb{R} \to \mathbb{R}$, whose spectrum is an infinite set, such that*

$$\langle \mathbf{u}, p \rangle = \int_{-\infty}^{+\infty} p(x) \, d\psi(x), \quad p \in \mathcal{P}. \tag{4.16}$$

Proof. We consider two cases.

Case 1. Assume that the true interval of orthogonality $[\xi, \eta]$ is bounded (compact). Then (cf. Remark 3.3) from (4.12) and (4.15) we see that $\psi(x) = 0$ if $x < \xi$, and $\psi(x) = u_0 > 0$ if $x > \eta$. Therefore, for each (fixed) $k \in \mathbb{N}_0$ we can write

$$\int_{-\infty}^{+\infty} x^k \, d\psi(x) = \int_{\xi}^{\eta} x^k \, d\psi(x) = \lim_{j \to \infty} \int_{\xi}^{\eta} x^k \, d\psi_{n_j}(x), \tag{4.17}$$

where the last equality holds by the Helly convergence theorem. Keeping k fixed, and since $k \leq 2n_j - 1$ for j sufficiently large, we deduce from (4.14) that the limit in (4.17) equals u_k, and so

$$\int_{-\infty}^{+\infty} x^k \, d\psi(x) = u_k = \langle \mathbf{u}, x^k \rangle, \quad k \in \mathbb{N}_0. \tag{4.18}$$

Thus, (4.16) follows whenever $[\xi, \eta]$ is bounded.

Case 2. Assume now that $[\xi, \eta]$ is unbounded. (In this case the Helly convergence theorem cannot be applied.) By the Helly selection theorem, there exists a subsequence $\{\psi_{n_j}\}_{j \geq 0}$ of $\{\psi_n\}_{n \geq 0}$ which converges on $\mathbb{R}$ to a bounded and nondecreasing function ψ. Setting $\phi_j := \psi_{n_j}$, according to (4.14) we deduce

$$\int_{-\infty}^{+\infty} x^k \, d\phi_j(x) = u_k \quad \text{if} \quad n_j \geq \frac{k+1}{2}, \quad k \in \mathbb{N}_0. \tag{4.19}$$

Fix $k \in \mathbb{N}_0$. For any compact interval $[\alpha, \beta]$, by the Helly convergence theorem we have

$$\lim_{j \to +\infty} \int_{\alpha}^{\beta} x^k \, d\phi_j(x) = \int_{\alpha}^{\beta} x^k \, d\psi(x). \tag{4.20}$$

Therefore, choosing $-\infty < \alpha < 0 < \beta < +\infty$ and j such that $n_j > k+1$, we deduce

$$\left| u_k - \int_\alpha^\beta x^k \, \mathrm{d}\psi(x) \right| = \left| \int_{-\infty}^{+\infty} x^k \, \mathrm{d}\phi_j(x) - \int_\alpha^\beta x^k \, \mathrm{d}\psi(x) \right| \tag{4.21}$$

$$= \left| \int_{-\infty}^{\alpha} x^k \, \mathrm{d}\phi_j(x) + \int_\alpha^\beta x^k \, \mathrm{d}\phi_j(x) + \int_\beta^{+\infty} x^k \, \mathrm{d}\phi_j(x) - \int_\alpha^\beta x^k \, \mathrm{d}\psi(x) \right|$$

$$\leq \left| \int_{-\infty}^{\alpha} x^k \, \mathrm{d}\phi_j(x) \right| + \left| \int_\beta^{+\infty} x^k \, \mathrm{d}\phi_j(x) \right| + \left| \int_\alpha^\beta x^k \, \mathrm{d}\phi_j(x) - \int_\alpha^\beta x^k \, \mathrm{d}\psi(x) \right| .$$

However,

$$\left| \int_{-\infty}^{\alpha} x^k \, \mathrm{d}\phi_j(x) \right| = \left| \int_{-\infty}^{\alpha} \frac{x^{2k+2}}{x^{k+2}} \, \mathrm{d}\phi_j(x) \right|$$

$$\leq \frac{1}{|\alpha|^{k+2}} \int_{-\infty}^{+\infty} x^{2k+2} \, \mathrm{d}\phi_j(x) = \frac{u_{2k+2}}{|\alpha|^{k+2}},$$

where the last equality follows from (4.19), since $n_j > k+1$. Similarly,

$$\left| \int_\beta^{+\infty} x^k \, \mathrm{d}\phi_j(x) \right| \leq \frac{u_{2k+2}}{\beta^{k+2}}.$$

Therefore, sending $j \to +\infty$ in (4.21) and taking into account (4.20), we find

$$\left| u_k - \int_\alpha^\beta x^k \, \mathrm{d}\psi(x) \right| \leq u_{2k+2} \left(\frac{1}{|\alpha|^{k+2}} + \frac{1}{\beta^{k+2}} \right).$$

Thus, taking the limits $\alpha \to -\infty$ and $\beta \to +\infty$, we conclude that (4.18) also holds whenever the true interval of orthogonality is unbounded. It remains to prove that the spectrum of any natural representative ψ (of $\mathbf{u}$) fulfilling (4.16) is an infinite set. Indeed, if $\sigma(\psi) = \{x_1, \cdots, x_N\}$ (a finite subset of $\mathbb{R}$), define

$$p(x) := (x - x_1)(x - x_2) \cdots (x - x_N),$$

and let $h_j := \psi(x_j + 0) - \psi(x_j - 0)$ be the jump of ψ at the point x_j. Then, from (4.16) we would have

$$\langle \mathbf{u}, p^2 \rangle = \int_{-\infty}^{+\infty} p^2(x) \, \mathrm{d}\psi(x) = \sum_{j=1}^{N} p^2(x_j) h_j = 0,$$

in contradiction with the positive definiteness of $\mathbf{u}$, and the proof is complete.

□

Remark 4.3. *We have remarked before that a natural representative* ψ *for a positive definite functional* $\mathbf{u} \in \mathcal{P}'$ *is a distribution function. Moreover, being a nondecreasing function, the set of points of discontinuity of* ψ *is finite or denumerable. Thus, since changing the values of* ψ *at its points of discontinuity does not change the value of the Riemann-Stieltjes integral with respect to* ψ *for continuous integrand functions (and so in particular for polynomials), it follows that there is a representative of* $\mathbf{u}$*, in the sense of* (4.16)*, which is a bounded nondecreasing right-continuous function with infinite spectrum and finite moments of all orders.*

Remark 4.4. *If* ψ *is a distribution function which represents a positive definite functional* $\mathbf{u} \in \mathcal{P}'$ *in the sense of (4.16), then so is any function obtained by adding a constant to* ψ*. Such distribution functions are called essentially equal. The discussion about the existence of different distribution functions (not essentially equal) which represent a given functional will be made later.*

4.3 The Spectral Theorem

In this section we present an alternative statement of the representation theorem (Theorem 4.4), called the *spectral theorem for orthogonal polynomials.* It is worth mentioning that both, the representation theorem and the spectral theorem, are equivalent versions of the Favard theorem in the positive definite case. The spectral theorem asserts that any positive definite functional $\mathbf{u} \in \mathcal{P}'$ admits an integral representation involving a positive Borel measure μ on $\mathbb{R}$ (which may not be unique) with infinite support and such that all its moments exist (i.e., they are finite). Recall that the support of μ is the set

$$\operatorname{supp}(\mu) := \big\{\mathrm{x} \in \mathbb{R} \,:\, \mu\big((\mathrm{x}-\epsilon, \mathrm{x}+\epsilon)\big) > 0,\ \epsilon > 0\big\}, \tag{4.22}$$

while saying that all the moments of μ exist (are finite) means that

$$\int |x|^n \, \mathrm{d}\mu < \infty, \quad n \in \mathbb{N}_0. \tag{4.23}$$

Given a finite positive Borel measure μ on $\mathbb{R}$, the function $F_\mu : \mathbb{R} \to \mathbb{R}$ defined by

$$F_\mu(x) := \mu\big((-\infty, x]\big) \tag{4.24}$$

is called the distribution function of μ. This function F_μ is bounded, nondecreasing, right-continuous, nonnegative, and it fulfills

$$\lim_{x \to -\infty} F_\mu(x) = 0.$$

Conversely, any function $F : \mathbb{R} \to \mathbb{R}$ satisfying these five properties is a distribution function for a finite positive Borel measure μ, so that $F \equiv F_\mu$, and

$$\int f(x)\, \mathrm{d}F_\mu(x) = \int f(x)\, \mathrm{d}\mu(x) \tag{4.25}$$

for each continuous and $\mu-$integrable function f, where the integral on the left-hand side of (4.25) is the Riemann-Stieltjes integral generated by F. Because of this fact, we will often use μ to denote both a measure and its corresponding distribution function. The integral on the left-hand side of (4.25) is indeed the Lebesgue-Stieltjes integral generated by F, and this is simply the Lebesgue integral with respect to the Lebesgue–Stieltjes measure μ_F generated by F.

Remark 4.5. *Notice that the support of a measure μ and the spectrum of the corresponding distribution function, F_μ, coincide, i.e.,*

$$\mathrm{supp}(\mu) = \sigma(\mathrm{F}_\mu) := \big\{\mathrm{x} \in \mathbb{R} \,:\, \mathrm{F}_\mu(\mathrm{x}+\delta) - \mathrm{F}_\mu(\mathrm{x}-\delta) > 0,\ \delta > 0\big\}. \tag{4.26}$$

Theorem 4.5 (Spectral Theorem). *Let $\{P_n\}_{n\geq 0}$ be a monic OPS characterized by the three-term recurrence relation*

$$P_{n+1}(x) = (x - \beta_n)P_n(x) - \gamma_n P_{n-1}(x), \tag{4.27}$$

with initial conditions $P_{-1}(x) = 0$ and $P_0(x) = 1$. Suppose that

$$\beta_{n-1} \in \mathbb{R}, \quad \gamma_n > 0, \quad n \in \mathbb{N}. \tag{4.28}$$

Then, there exists a positive Borel measure μ on $\mathbb{R}$, whose support is an infinite set, and with finite moments of all orders, such that

$$\int P_n(x)P_m(x)\, \mathrm{d}\mu(x) = \zeta_n \delta_{n,m}, \quad m \in \mathbb{N}_0, \tag{4.29}$$

where $\zeta_0 := 1$ and $\zeta_n := \gamma_1\gamma_2\cdots\gamma_n$.

Proof. By the Favard theorem, under the given hypothesis $\{P_n\}_{n\geq 0}$ is a monic OPS with respect to a positive definite functional $\mathbf{u} \in \mathcal{P}'$. Therefore, by the representation Theorem 4.4, there exists a distribution function $\psi : \mathbb{R} \to \mathbb{R}$ (a natural representative of $\mathbf{u}$), whose spectrum $\sigma(\psi)$ is an infinite subset of $\mathbb{R}$, fulfilling

$$u_k := \langle \mathbf{u}, x^k \rangle = \int_{-\infty}^{+\infty} x^k \, \mathrm{d}\psi(x), \quad k \in \mathbb{N}_0. \tag{4.30}$$

We may assume that ψ is right-continuous without changing its spectrum. Let μ be the corresponding Stieltjes-Lebesgue measure (hence it is a positive Borel measure), so that ψ is the distribution function F_μ of the measure μ. Then

$$\psi(x) = F_\mu(x) = \mu\big((-\infty, x]\big), \quad x \in \mathbb{R},$$

and we get $\operatorname{supp}(\mu) = \sigma(\psi)$. Therefore, the support of μ is an infinite set. Moreover, from the connection between the Riemann-Lebesgue and the Stieltjes-Lebesgue integrals, we have

$$\int_{-\infty}^{+\infty} p(x)\,\mathrm{d}\psi(x) = \int p(x)\,\mathrm{d}F_\mu(x) = \int p(x)\,\mathrm{d}\mu(x), \quad p \in \mathcal{P},$$

and so, in particular, by (4.30), the moments of μ all exist and

$$\begin{aligned}\int P_n(x)P_m(x)\,\mathrm{d}\mu(x) &= \int_{-\infty}^{+\infty} P_n(x)P_m(x)\,\mathrm{d}\psi(x)\\ &= \langle \mathbf{u}, P_nP_m\rangle = \langle \mathbf{u}, P_n^2\rangle \delta_{n,m}, \quad m \in \mathbb{N}_0.\end{aligned}$$

Now, taking into account (2.29), we have

$$\langle \mathbf{u}, P_n^2\rangle = u_0 \prod_{j=1}^{n} \gamma_j = u_0 \zeta_n, \quad n \in \mathbb{N}_0.$$

Thus, if $u_0 = 1$, we obtain (4.29); otherwise, we normalize μ passing to the measure $\widehat{\mu} = u_0^{-1}\mu$, and so (4.29) holds with $\widehat{\mu}$ instead of μ. □

Often, we will refer to a measure μ under the conditions of the spectral theorem as a "spectral measure", or an "orthogonality measure" for the given OPS $\{P_n\}_{n\geq 0}$. This measure does not need to be unique whenever the true interval of orthogonality is an unbounded set, as was observed by Stieltjes. In the next section we consider this question.

4.4 The Unicity of the Spectral Measure

As remarked before, the orthogonality measure for an OPS does not need to be unique. We present an example due to Stieltjes. Consider the weight function

$$w_c(x;\alpha) := \big(1+\alpha\,\sin(2\pi c\ln x)\big)e^{-c\ln^2 x}, \quad x\in I := (0,+\infty), \tag{4.31}$$

where we fix α and c so that $-1<\alpha<1$ and $c>0$. Clearly, these choices of c and α ensure that $w_c(x;\alpha)$ becomes nonnegative and integrable on I. This weight function defines a positive definite functional $\mathbf{u}\in\mathcal{P}'$ given by

$$\langle \mathbf{u}, p\rangle := \int_0^{+\infty} p(x)w_c(x;\alpha)\,\mathrm{d}x, \quad p\in\mathcal{P}.$$

An associated distribution function representing $\mathbf{u}$ is

$$\psi_c(x;\alpha) := \int_{-\infty}^{x} w_c(t;\alpha)\chi_{(0,+\infty)}(t)\,\mathrm{d}t, \quad x\in\mathbb{R}.$$

Computing the moments of $\mathbf{u}$, we deduce

$$u_n := \int_0^{+\infty} x^n w_c(x;\alpha)\,\mathrm{d}x = \sqrt{\frac{\pi}{c}}\,e^{(n+1)^2/4c}, \quad n\in\mathbb{N}_0. \tag{4.32}$$

Thus, we see that the moments are independent of the choice of α, hence (for fixed $c>0$) by varying $\alpha\in(-1,1)$ we obtain different orthogonality measures with the same moments. Therefore there are infinitely many orthogonality measures for an OPS with respect to $\mathbf{u}$. Next, we prove that the orthogonality measure given by the spectral theorem is unique if both sequences of the $\beta-$parameters and $\gamma-$parameters are bounded. We begin by stating two preliminary results.

Lemma 4.3. *Let $A=[a_{ij}]_{i,j=1}^N$ be a matrix of order N, and let M be a positive constant chosen so that*

$$|a_{i,j}|\le M, \quad i,j=1,\dots,N.$$

Suppose that each row and each column of A has at most ℓ nonzero entries. Then each eigenvalue λ of A satisfies

$$|\lambda|\le \ell M$$

Proof. Take $x=[x_1\,\dots\,x_N]^T$ to be an eigenvector of A corresponding to the eigenvalue λ, so that

$$Ax=\lambda x, \quad \|x\|=1,$$

where $\|x\| := \sqrt{\langle x,x\rangle}$, being $\langle x,y\rangle := \sum_{j=1}^N x_j\overline{y_j}$ the usual inner product on

$\mathbb{C}^N$. Then

$$|\lambda|^2 = \left|\lambda\,\|x\|^2\right|^2 = \left|\lambda\langle x,x\rangle\right|^2 = \left|\langle \lambda x,x\rangle\right|^2 = \left|\langle Ax,x\rangle\right|^2$$
$$\leq \|Ax\|^2\,\|x\|^2 = \|Ax\|^2 = \sum_{j=1}^{N}|(Ax)_j|^2 = \sum_{j=1}^{N}\Big|\sum_{k=1}^{N} a_{jk}x_k\Big|^2$$
$$\leq \sum_{j=1}^{N}\Big(\sum_{k=1}^{N}|a_{jk}|^2\sum_{k=1}^{N}|x_k|^2\Big) = \sum_{j=1}^{N}\sum_{k=1}^{N}|a_{jk}|^2 \leq \ell^2 M^2,$$

where in the first two inequalities we have applied the Cauchy-Schwartz inequality, and the last one holds since, by hypothesis, each row and each column of A has at most ℓ nonzero entries and all the entries of A are bounded by M. □

Theorem 4.6. *Under the hypothesis of Theorem 4.5, assume further that both $\{\beta_n\}_{n\geq 0}$ and $\{\gamma_n\}_{n\geq 1}$ are bounded sequences. Then the support of the orthogonality measure μ is a bounded set.*

Proof. By hypothesis,

$$\exists C > 0 \,:\, \forall n \in \mathbb{N}, \quad |\beta_n| \leq C, \quad |\gamma_n| \leq C. \tag{4.33}$$

On the other hand, we know that the zeros $x_{n,j}$ of P_n are the eigenvalues of the tridiagonal matrix J_n given by (3.17). Since, by (4.33), the entries of J_n are bounded by $M := \max\{1, C\}$, and in each row and each column of J_n there are at most $\ell = 3$ nonzero entries, then Lemma 4.3 ensures that

$$\left|x_{n,j}\right| \leq 3M, \quad j = 1, \ldots, n. \tag{4.34}$$

Therefore, the spectrum of each distribution function $\psi_n : \mathbb{R} \to \mathbb{R}$ introduced in (4.12) is contained in the interval $[-3M, 3M]$, hence the spectrum $\sigma(\psi)$ of any distribution function ψ obtained as a limit of a subsequence of $\{\psi_n\}_{n\geq 1}$ is also contained in $[-3M, 3M]$. Thus, the orthogonality measure μ given by the spectral theorem satisfies

$$\operatorname{supp}(\mu) = \sigma(\psi) \subseteq [-3M, 3M],$$

so that $\operatorname{supp}(\mu)$ is a bounded set, which proves the theorem. □

Remark 4.6. *Under the conditions of the spectral theorem, it follows from the proof of Theorem 4.6 that if $\{\beta_n\}_{n\geq 0}$ and $\{\gamma_n\}_{n\geq 1}$ are bounded sequences then the true interval of orthogonality of the corresponding monic OPS is bounded.*

Theorem 4.7. *Under the hypothesis of Theorem 4.5, assume further that both $\{\beta_n\}_{n\geq 0}$ and $\{\gamma_n\}_{n\geq 1}$ are bounded sequences. Then the orthogonality measure μ is unique.*

Proof. By Theorem 4.6, there exists at least one orthogonality measure μ with compact support. Let ν be any other orthogonality measure (hence it has the same moments as μ). For $a > 0$ fixed, we have

$$\begin{aligned}\int_{|x|\geq a} \mathrm{d}\nu(x) &\leq \int_{|x|\geq a} \left|\frac{x}{a}\right|^{2n} \mathrm{d}\nu(x) \leq a^{-2n} \int x^{2n}\, \mathrm{d}\nu(x)\\ &= a^{-2n} \int x^{2n}\, \mathrm{d}\mu(x), \quad n \in \mathbb{N}_0. \end{aligned} \tag{4.35}$$

We have seen in the proof of Theorem 4.6 that the zeros $x_{n,j}$ of P_n are uniformly bounded, hence

$$\exists\, r > 0 \;:\; \forall\, n \in \mathbb{N}, \quad j \in \{1, 2, \ldots, n+1\}, \quad |x_{n+1,j}| \leq r. \tag{4.36}$$

By the Gauss-Jacobi mechanical quadrature formula (Theorem 3.4), with n replaced by $n+1$ and $p(x) = x^{2n}$, we have

$$\int x^{2n}\, \mathrm{d}\mu(x) = \sum_{j=1}^{n+1} A_{n+1,j} x_{n+1,j}^{2n}.$$

Substituting this in the right-hand side of (4.35) and taking into account (4.36), as well as (3.23), we deduce

$$\int_{|x|\geq a} \mathrm{d}\nu(x) \leq \left(\frac{r}{a}\right)^{2n} \sum_{j=1}^{n+1} A_{n+1,j} = u_0 \left(\frac{r}{a}\right)^{2n}.$$

Choosing $a > r$ and taking the limit as $n \to +\infty$, we obtain

$$\int_{|x|\geq a} \mathrm{d}\nu(x) = 0.$$

Thus $\nu\big(\{x \in \mathbb{R} : |x| \geq a\}\big) = 0$, hence $\operatorname{supp}(\nu) \subseteq [-a, a]$ for each $a > r$, and so

$$\operatorname{supp}(\nu) \subseteq [-r, r].$$

This proves that any orthogonality measure has a compact support contained in the interval $[-r, r]$. To prove the uniqueness of the orthogonality measure, take arbitrarily $z \in \mathbb{C}$ and $t \in \mathbb{R}$ such that $|z| \geq 2r$ and $|t| \leq r$. Since $|t/z| \leq 1/2$,

$$s_n(t) := \sum_{k=0}^{n} \frac{t^k}{z^{k+1}} = \frac{1}{z}\sum_{k=0}^{n}\left(\frac{t}{z}\right)^k = \frac{1-\left(\frac{t}{z}\right)^{n+1}}{z-t} \to \frac{1}{z-t} \quad \text{as } n \to +\infty.$$

Therefore, for any orthogonality measure μ,

$$\int \frac{\mathrm{d}\mu(t)}{z-t} = \int \lim_{n\to\infty} s_n(t)\,\mathrm{d}\mu(t) = \lim_{n\to\infty}\int s_n(t)\,\mathrm{d}\mu(t) = \lim_{n\to\infty}\sum_{k=0}^{n}\frac{u_k}{z^{k+1}}, \tag{4.37}$$

where the interchange between the limit and the integral follows by the Lebesgue dominated convergence theorem, taking into account that

$$|s_n(t)| = \left|\frac{1-\left(\frac{t}{z}\right)^{n+1}}{z-t}\right| \leq \frac{2}{|z-t|} =: g(t) \in L^1_\mu([-r,r]).$$

Notice that the limit on the right-hand side of (4.37) depends only on the moments u_k, hence it has the same value considering any measure μ with the same moments and with compact support contained in $[-r,r]$ (i.e., considering any orthogonality measure). Therefore, the function

$$F(z) := \int \frac{\mathrm{d}\mu(x)}{z-x}, \quad z \in \mathbb{C}\setminus[-r,r]$$

is uniquely determined by μ for z outside the circle $|z| = 2r$ (meaning that, for z outside this circle, $F(z)$ has the same value for any orthogonality measure μ). Since F is analytic on $\mathbb{C}\setminus[-r,r]$, then by the identity theorem for analytic functions, we conclude that $F(z)$ is uniquely determined by μ for $z \in \mathbb{C}\setminus[-r,r]$. Thus the uniqueness of the measure follows from the Perron-Stieltjes inversion formula:

$$\psi_\mu(t) - \psi_\mu(s) = \lim_{\varepsilon\to 0^+}\frac{1}{\pi}\int_s^t \frac{F(x-i\varepsilon)-F(x+i\varepsilon)}{2i}\,\mathrm{d}x,$$

where ψ_μ is an appropriate normalization of the distribution function F_μ, given by

$$\psi_\mu(-\infty) := 0, \quad \psi_\mu(t) := \frac{F_\mu(t+0)+F_\mu(t-0)}{2}, \quad t \in \mathbb{R}.$$

(The Perron–Stieltjes inversion formula will be proved later.) Notice that ψ_μ and F_μ may be different only at (countably many) points of discontinuity, and so the integrals of continuous functions with respect to ψ_μ and F_μ take the same value. □

We conclude this section by stating without proof two results that ensure the uniqueness of the orthogonality measure—see Theorems II-5.1 and II-5.2 in the Freud's book [18].

Theorem 4.8 (Riesz Uniqueness Criterium). *The orthogonality measure μ is unique whenever its sequence of moments*

$$u_n := \int_{\mathbb{R}} t^n \, \mathrm{d}\mu(t), \quad n \in \mathbb{N}_0,$$

satisfies

$$\liminf_{n\to+\infty} \frac{\sqrt[2n]{u_{2n}}}{2n} < \infty. \tag{4.38}$$

Corollary 4.1. *The orthogonality measure μ is unique if the condition*

$$\int_{\mathbb{R}} e^{\theta|x|} \, \mathrm{d}\mu(x) < \infty \tag{4.39}$$

holds for some $\theta > 0$.

Exercises

1. Show that the conclusion of the Helly convergence theorem (Theorem 4.2) may not hold whenever the true interval of orthogonality is not a bounded interval.
2. Prove (4.32).
3. (Charlier polynomials.) Define the monic Charlier OPS $\{C_n^{(a)}(x)\}_{n\geq 0}$ by the generating function

$$e^{-aw}(1+w)^x = \sum_{n=0}^{\infty} C_n^{(a)}(x) \frac{w^n}{n!}, \quad a \in \mathbb{R}\setminus\{0\}.$$

Prove the following assertions:

(a) $C_n^{(a)}(x)$ has the explicit representation

$$C_n^{(a)}(x) = \sum_{k=0}^{n} \binom{x}{k}\binom{n}{k} k!(-a)^{n-k},$$

where

$$\binom{z}{0} := 1, \quad \binom{z}{k} := z(z-1)\cdots(z-k+1)/k!, \quad k = 1, 2, \dots, n, \ z \in \mathbb{C}.$$

(b) $\{C_n^{(a)}(x)\}_{n\geq 0}$ is an OPS with respect to the functional $\mathbf{u} \in \mathcal{P}'$ given by

$$\langle \mathbf{u}, p \rangle = \int_0^{+\infty} p(x)\, \mathrm{d}\psi^{(a)}(x), \quad p \in \mathcal{P},$$

where $\psi : \mathbb{R} \to \mathbb{R}$ is a step function whose jumps are given by $e^{-a}a^x/x!$ at the points $x = 0, 1, 2, \dots$. The positive definite case occurs for $a > 0$, and in this case $\psi^{(a)}(x)$ is the Poisson distribution function of probability theory.

(c) The three-term recurrence relation for $\{C_n^{(a)}(x)\}_{n\geq 0}$ is

$$C_{n+1}^{(a)}(x) = (x - n - a)C_n^{(a)}(x) - anC_{n-1}^{(a)}(x),$$

where $C_{-1}^{(a)}(x) := 0$ and $C_0^{(a)}(x) = 0$.

4. (Meixner polynomials.) Let $\{m_n(x;\beta,c)\}_{n\geq 0}$ be the Meixner OPS of the first kind, defined via the generating function

$$\left(1 - c^{-1}w\right)^x (1 - w)^{-x-\beta} = \sum_{n=0}^{\infty} m_n(x;\beta,c)\, \frac{w^n}{n!},$$

being $c \in \mathbb{R} \setminus \{0, 1\}$ and $\beta \in \mathbb{R} \setminus \{0, -1, -2, -3, \dots\}$. Prove the following assertions:

(a) $m_n(x;\beta,c)$ has the explicit representation

$$m_n(x;\beta,c) = (-1)^n n! \sum_{k=0}^{n} \binom{x}{k} \binom{-x-\beta}{n-k} c^{-k}.$$

(b) If $0 < c < 1$ and $\beta > 0$, $\{m_n(x;\beta,c)\}_{n\geq 0}$ is an OPS with respect to the positive definite functional $\mathbf{u} \in \mathcal{P}'$ given by

$$\langle \mathbf{u}, p \rangle = \int_{-\infty}^{+\infty} p(x)\, \mathrm{d}\psi(x), \quad p \in \mathcal{P},$$

where the distribution function $\psi : \mathbb{R} \to \mathbb{R}$ is a step function supported on $\mathbb{N}_0$ (i.e., $\sigma(\psi) = \mathbb{N}_0$) whose jump at the point $x = k$ is given by $c^k(\beta)_k/k!$, $k \in \mathbb{N}_0$.

5. Let $\{P_n\}_{n\geq 0}$ be a monic OPS with respect to a positive Borel measure μ. Prove that each $P_n(x)$ admits the representation

$$P_n(x) = \frac{1}{n!H_{n-1}} \int \cdots \int \prod_{i=1}^{n} (x - x_j) \prod_{1\leq j<k\leq n} (x_j - x_k)^2\, \mathrm{d}\mu(x_1)\cdots \mathrm{d}\mu(x_n),$$

where H_{n-1} is the Hankel determinant of order n.

Chapter 5

The Markov Theorem

In this chapter we describe a method that leads to the orthogonality measure from the three-term recurrence relation, whose main tools are the Markov theorem and the Perron-Stieltjes inversion formula. The latter allows us to find the measure from the Stieltjes transform, which in turn is determined by the former.

5.1 The Perron-Stieltjes Inversion Formula

According to Theorem 4.5, given a sequence of monic polynomials, $\{P_n\}_{n\geq 0}$, satisfying the three-term recurrence relation

$$xP_n(x) = P_{n+1}(x) + \beta_n P_n(x) + \gamma_n P_{n-1}(x), \tag{5.1}$$

with initial conditions $P_{-1}(x) = 0$ and $P_0(x) = 1$, and subject to the conditions

$$\beta_{n-1} \in \mathbb{R}, \quad \gamma_n > 0, \tag{5.2}$$

there exists a positive Borel measure μ with respect to which $\{P_n\}_{n\geq 0}$ is an OPS. Moreover, when the true interval of orthogonality is a compact set (or when both $\{\beta_n\}_{n\geq 0}$ and $\{\gamma_n\}_{n\geq 1}$ are bounded sequences) this measure is unique. The natural question arises: How is it possible to find the orthogonality measure from the three-term recurrence relation (5.1)? This question fits into the study of the so-called inverse problems in the theory of OP.

Definition 5.1. *Let μ be a finite positive Borel measure with* $\mathrm{supp}(\mu) \subseteq \mathbb{R}$. *The Stieltjes transform associated with μ is the complex function $F \equiv F(\cdot;\mu)$ given by*

$$F(z) := \int \frac{\mathrm{d}\mu(x)}{z-x}, \quad z \in \mathbb{C} \setminus \mathrm{supp}(\mu). \tag{5.3}$$

DOI: 10.1201/9781003432067-5

Recall that $\mathrm{supp}(\mu)$ is a closed set. From this we can prove that F is an analytic function on $\mathbb{C}\setminus\mathrm{supp}(\mu)$. The Perron-Stieltjes inversion formula allows us to recover the measure from the corresponding Stieltjes transform.

Theorem 5.1 (Perron-Stieltjes Inversion Formula). *Let μ be a finite positive Borel measure. Then, for every $a, b \in \mathbb{R}$, with $a < b$, the equality*

$$\lim_{\varepsilon\to 0^+} \frac{1}{\pi}\int_a^b \Im\,(F(x-i\varepsilon))\,dx = \mu((a,b)) + \tfrac{1}{2}\,\mu(\{a\}) + \tfrac{1}{2}\,\mu(\{b\}) \tag{5.4}$$

holds, F being the Stieltjes transform associated with μ.

Proof. Observe that $\overline{F(z)} = F(\overline{z})$. Hence

$$\Im(F(z)) = \frac{F(z)-F(\overline{z})}{2i} = -\frac{1}{2i}\int \frac{z-\overline{z}}{|x-z|^2}\,\mathrm{d}\mu(x) = -\int \frac{\Im(z)}{|x-z|^2}\,\mathrm{d}\mu(x).$$

Setting $z = x - i\epsilon$, with $x \in \mathbb{R}$ and $\varepsilon > 0$, we may write

$$\Im(F(x-i\varepsilon)) = \int \frac{\varepsilon}{|s-x+i\varepsilon|^2}\,\mathrm{d}\mu(s) = \int \frac{\varepsilon}{(s-x)^2+\varepsilon^2}\,\mathrm{d}\mu(s).$$

Integrating on (a,b) with respect to x, and then interchanging the order of integration in the last integral (this is allowed taking into account that the integrand function is positive), we obtain

$$\int_a^b \Im(F(x-i\varepsilon))\,\mathrm{d}x = \int \theta_\varepsilon(s)\,\mathrm{d}\mu(s), \tag{5.5}$$

where

$$\theta_\varepsilon(s) := \int_a^b \frac{\varepsilon}{(s-x)^2+\varepsilon^2}\,\mathrm{d}x = \arctan\left(\frac{b-s}{\varepsilon}\right) - \arctan\left(\frac{a-s}{\varepsilon}\right).$$

Note that, for each $s \in \mathbb{R}$,

$$\lim_{\varepsilon\to 0}\theta_\varepsilon(s) = \begin{cases} \pi, & a < s < b \\ \dfrac{\pi}{2}, & s = a \text{ or } s = b \\ 0, & s < a \text{ or } s > b \end{cases}$$

$$= \pi\chi_{(a,b)}(s) + \frac{\pi}{2}\,\chi_{\{a\}}(s) + \frac{\pi}{2}\,\chi_{\{b\}}(s).$$

Moreover, $|\theta_\varepsilon(s)| \le \pi$ for each $s \in \mathbb{R}$, and the (constant) function $s \in \mathbb{R} \mapsto \pi$ is integrable with respect to μ, because μ is a finite measure, so that $\mu(\mathbb{R}) < \infty$. Hence, by the Lebesgue dominated convergence theorem,

$$\begin{aligned}\lim_{\varepsilon\to 0}\int \theta_\varepsilon(s)\,\mathrm{d}\mu(s) &= \int \lim_{\varepsilon\to 0}\theta_\varepsilon(s)\,\mathrm{d}\mu(s)\\ &= \pi\mu((a,b)) + \frac{\pi}{2}\,\mu(\{a\}) + \frac{\pi}{2}\,\mu(\{b\}),\end{aligned}$$

and the theorem follows from (5.5) taking the limit as $\varepsilon \to 0^+$. □

The proof of Theorem 5.1 is taken from [25].

Remark 5.1. *The Perron-Stieltjes inversion formula (5.4) may be stated in terms of the distribution function F_μ associated to the measure μ, after an appropriate normalization of F_μ. Indeed, being $\psi : \mathbb{R} \to \mathbb{R}$ defined by*

$$\psi(x) := \frac{F_\mu(x+0) + F_\mu(x-0)}{2},$$

and recalling that $F_\mu(x) := \mu((-\infty, x])$ for each $x \in \mathbb{R}$, (5.4) may be rewritten as

$$\psi(b) - \psi(a) = \lim_{\varepsilon\to 0^+} \frac{1}{\pi}\int_a^b \Im(F(x - i\varepsilon))\,\mathrm{d}x.$$

In fact, we deduce

$$\begin{aligned}\psi(b) - \psi(a) &= \frac{1}{2}\,(F_\mu(b+0) + F_\mu(b-0) - F_\mu(a+0) - F_\mu(a-0))\\ &= \frac{1}{2}\,((F_\mu(b) - F_\mu(a)) + (F_\mu(b-0) - F_\mu(a-0)))\\ &= \frac{1}{2}\,(\mu((a,b]) + \mu([a,b))) = \mu((a,b)) + \frac{1}{2}\mu(\{a\}) + \frac{1}{2}\mu(\{b\}),\end{aligned}$$

where in the second equality we took into account that F_μ is right-continuous.

5.2 Associated Polynomials

Let $\mathbf{u} \in \mathcal{P}'$ be regular (not necessarily positive definite), and let $\{P_n\}_{n\geq 0}$ be the corresponding monic OPS. According to the Favard theorem, $\{P_n\}_{n\geq 0}$ is characterized by the three-term recurrence relation (5.1), with $\beta_{n-1} \in \mathbb{C}$ and $\gamma_n \in \mathbb{C} \setminus \{0\}$. Making a shift on this recurrence relation, we may define a new monic OPS, $\{P_n^{(k)}\}_{n\geq 0}$, where we have set $k \in \mathbb{N}_0$, by

$$P_{n+1}^{(k)}(x) = (x - \beta_{n+k})\, P_n^{(k)}(x) - \gamma_{n+k} P_{n-1}^{(k)}(x), \tag{5.6}$$

with initial conditions $P_{-1}^{(k)}(x) = 0$ and $P_0^{(k)}(x) = 1$. The Favard theorem ensures that, indeed, $\{P_n^{(k)}\}_{n\geq 0}$ is a monic OPS.

Remark 5.2. *If $k = 0$, then $P_n^{(0)} \equiv P_n$. When $k = 1$, often $\{P_n^{(1)}\}_{n\geq 0}$ is called the sequence of (monic) associated polynomials of the first kind, or numerator polynomials.*

According with (3.18), $P_n^{(k)}$ has the following representation as a determinant of order n of a tridiagonal matrix:

$$P_n^{(k)}(x) = \begin{vmatrix} x-\beta_k & 1 & 0 & \cdots & 0 & 0 \\ \gamma_{k+1} & x-\beta_{k+1} & 1 & \cdots & 0 & 0 \\ 0 & \gamma_{k+2} & x-\beta_{k+2} & \cdots & 0 & 0 \\ \vdots & \vdots & \vdots & \ddots & \vdots & \vdots \\ 0 & 0 & 0 & \cdots & x-\beta_{n+k-2} & 1 \\ 0 & 0 & 0 & \cdots & \gamma_{n+k-1} & x-\beta_{n+k-1} \end{vmatrix}. \tag{5.7}$$

When $k \in \mathbb{N}$, another representation for $P_n^{(k)}$ is

$$P_n^{(k)}(x) = \frac{1}{\langle \mathbf{u}, P_{k-1}^2\rangle} \left\langle P_{k-1}(\xi)\mathbf{u}(\xi), \frac{P_{n+k}(x) - P_{n+k}(\xi)}{x-\xi} \right\rangle, \tag{5.8}$$

where $\mathbf{u}(\xi)$ means that $\mathbf{u}$ acts on polynomials regarded as functions of the variable ξ. The representation (5.8) may be easily proved by checking that the right-hand side defines a polynomial on the variable x which fulfills the three-term recurrence relation (5.6), and for $n = -1$ and $n = 0$ the right-hand side of (5.8) equals 0 and 1, respectively. A more concise form of writing (5.8) is

$$P_n^{(k)}(x) = \langle (\xi - x)^{-1}\mathbf{a}_{k-1}(\xi), P_{n+k}(\xi)\rangle\,, \tag{5.9}$$

where $\{\mathbf{a}_n\}_{n\geq 0}$ is the dual basis associated with $\{P_n\}_{n\geq 0}$ and the division of a functional by a polynomial is given by Definition 1.1. Finally, considering the operator θ_0 introduced in (1.8) and taking into account the definition of the right multiplication of a functional by a polynomial (Definition 1.4), then from (5.9) we arrive at a rather elegant representation of the nth degree monic associated polynomial of order k:

$$P_n^{(k)} = \mathbf{a}_{k-1}\theta_0 P_{n+k}. \tag{5.10}$$

A very useful relation linking the associated polynomials of orders k and $k+1$ is

$$P_n^{(k+1)}(x)P_n^{(k)}(x) - P_{n+1}^{(k)}(x)P_{n-1}^{(k+1)}(x) = \prod_{j=1}^{n} \gamma_{j+k}. \tag{5.11}$$

Indeed, by (5.6), we have

$$P_n^{(k+1)}(x) = (x - \beta_{n+k})\, P_{n-1}^{(k+1)}(x) - \gamma_{n+k} P_{n-2}^{(k+1)}(x),$$

$$P_{n+1}^{(k)}(x) = (x - \beta_{n+k})\, P_n^{(k)}(x) - \gamma_{n+k} P_{n-1}^{(k)}(x).$$

Multiplying the first equality by $P_n^{(k)}(x)$ and the second one by $-P_{n-1}^{(k+1)}(x)$, and then adding the resulting equalities, we deduce

$$\begin{aligned} &P_n^{(k+1)}(x)P_n^{(k)}(x) - P_{n+1}^{(k)}(x)P_{n-1}^{(k+1)}(x) \\ &\quad = \gamma_{n+k}\left(P_{n-1}^{(k+1)}(x)P_{n-1}^{(k)}(x) - P_n^{(k)}(x)P_{n-2}^{(k+1)}(x)\right), \end{aligned}$$

and so (5.11) follows after a repeated application of this relation. We also state the following formulas, close to the Christoffel-Darboux identities (Theorem 2.10):

$$\frac{P_n(x) - P_n(y)}{x - y} = \sum_{k=1}^{n} P_{k-1}(x)P_{n-k}^{(k)}(y), \quad n \in \mathbb{N}, \tag{5.12}$$

$$P_n'(x) = \sum_{k=1}^{n} P_{k-1}(x)P_{n-k}^{(k)}(x), \quad n \in \mathbb{N}. \tag{5.13}$$

Clearly, (5.13) follows from (5.12) by taking the limit $y \to x$. To prove (5.12), notice that $(P_n(x) - P_n(y))/(x - y)$ is a polynomial of degree $n - 1$ in the variable x (whose coefficients depend on y), so we can write

$$\frac{P_n(x) - P_n(y)}{x - y} = \sum_{k=0}^{n-1} c_{n,k}(y)P_k(x). \tag{5.14}$$

Then for $0 \le k \le n - 1$, we compute the Fourier coefficients $c_{n,k}(y)$:

$$c_{n,k}(y) = \frac{\left\langle \mathbf{u}(x), \dfrac{P_n(x) - P_n(y)}{x - y} P_k(x) \right\rangle}{\langle \mathbf{u}, P_k^2 \rangle} = P_{n-1-k}^{(k+1)}(y),$$

where the last equality holds due to (5.8). Substituting the last expression for $c_{n,k}(y)$ into (5.14) we obtain (5.12).

5.3 The Markov Theorem

We now return to the positive definite case, with the purpose of stating the celebrated Markov theorem. We begin by proving some preliminary results.

Lemma 5.1. *Let* $\mathbf{u} \in \mathcal{P}'$ *be positive definite and let* $\{P_n\}_{n\geq 0}$ *be the corresponding monic OPS. Then*

$$\frac{u_0 P_{n-1}^{(1)}(x)}{P_n(x)} = \sum_{j=1}^{n} \frac{A_{n,j}}{x - x_{n,j}} \tag{5.15}$$

$$= \int_{-\infty}^{+\infty} \frac{\mathrm{d}\psi_n(t)}{x-t}, \quad x \in \mathbb{C} \setminus \Lambda_n,$$

where $A_{n,1}, \ldots, A_{n,n}$ *are the coefficients appearing in the quadrature formula* (3.23), $\Lambda_n := \{x_{n,1}, \ldots, x_{n,n}\}$, *being* $x_{n,1} < \cdots < x_{n,n}$ *the zeros of* P_n, *and* ψ_n *is the distribution function introduced in* (4.12).

Proof. Notice first that, by (5.11) with $k = 0$, for each $n \in \mathbb{N}$ the equality

$$P_n^{(1)}(x)P_n(x) - P_{n+1}(x)P_{n-1}^{(1)}(x) = \gamma_1\gamma_2\cdots\gamma_n$$

holds. Hence, since $\gamma_1\gamma_2\cdots\gamma_n \neq 0$, we see that P_n and $P_{n-1}^{(1)}$ have no common zeros. Moreover, we know that the zeros of P_n are real and simple. Thus, for each $n \in \mathbb{N}$, the decomposition of the rational function $P_{n-1}^{(1)}(x)/P_n(x)$ into partial fractions yields

$$\frac{P_{n-1}^{(1)}(x)}{P_n(x)} = \sum_{j=1}^{n} \frac{\lambda_{n,j}}{x - x_{n,j}}, \quad \lambda_{n,j} := \frac{P_{n-1}^{(1)}(x_{n,j})}{P_n'(x_{n,j})}. \tag{5.16}$$

On the other hand, from (3.25) in the proof of Theorem 3.4, we know that

$$A_{n,j} = \langle \mathbf{u}, \ell_{j,n} \rangle = \left\langle \mathbf{u}, \frac{P_n(x)}{(x - x_{n,j})P_n'(x_{n,j})} \right\rangle$$

$$= \frac{u_0}{P_n'(x_{n,j})} \frac{1}{u_0} \left\langle \mathbf{u}, \frac{P_n(x_{n,j}) - P_n(x)}{x_{n,j} - x} \right\rangle.$$

Moreover, by (5.8) with $k = 1$, and changing n into $n-1$, we see that the relation

$$P_{n-1}^{(1)}(x) = \frac{1}{u_0} \left\langle \mathbf{u}_\xi, \frac{P_n(x) - P_n(\xi)}{x - \xi} \right\rangle$$

holds for each $n \in \mathbb{N}$. Therefore, we conclude that

$$A_{n,j} = u_0 P_{n-1}^{(1)}(x_{n,j})/P_n'(x_{n,j}).$$

Hence $\lambda_{n,j} = A_{n,j}/u_0$. Thus the first equality in (5.15) is proved. The second one is an immediate consequence of the properties of the Riemann-Stieltjes integral, taking into account that ψ_n is a step function with spectrum $\sigma(\psi_n) = \Lambda_n$, and with jump equal to $A_{n,j}$ at the point $x_{n,j}$. □

Lemma 5.2. *Let $\{P_n\}_{n\geq 0}$ be a monic OPS with respect to a positive definite functional $\mathbf{u} \in \mathcal{P}'$. Let ψ be a natural representative for $\mathbf{u}$, in the sense of Theorem 4.4. Let s be a spectral point of ψ. Then, every neighborhood of s contains at least one zero of $P_n(x)$ for infinitely many values of $n \in \mathbb{N}$. That is*

$$\forall V \in \mathcal{V}(s),\ \forall N \in \mathbb{N},\ \exists n \in \mathbb{N} \ :\ n > N \ \wedge \ V \cap \Lambda_n \neq \emptyset, \tag{5.17}$$

$\mathcal{V}(s)$ being the set of all neighborhoods of s.

Proof. By hypothesis, $s \in \sigma(\psi)$. If (5.17) is not true, there exists a neighborhood V of s and $N \in \mathbb{N}$ such that V does not contain zeros of P_n for every $n \geq N$. Then, by definition of ψ_n, we have $\psi_n(x) = \psi_n(z)$ for all $x, z \in V$ and $n \geq N$. Since ψ is a natural representative of $\mathbf{u}$, there is a subsequence $\{\psi_{n_j}\}_{j\geq 1}$ of $\{\psi_n\}_{n\geq 1}$ such that

$$\psi(x) := \lim_{j\to\infty} \psi_{n_j}(x), \quad x \in \mathbb{R}.$$

Therefore, $\psi(x) = \lim_{j\to\infty} \psi_{n_j}(x) = \lim_{j\to\infty} \psi_{n_j}(z) = \psi(z)$ for every $x, z \in V$. Thus ψ is constant on the neighborhood V of s, hence $s \notin \sigma(\psi)$, contrary to the hypothesis. □

It follows from Lemma 5.2 that if s is a spectral point of a natural representative ψ of $\mathbf{u}$, then either s is a zero of $P_n(x)$ for infinitely many n, or else s is a limit of a sequence of numbers belonging to the set

$$Z_1 := \{x_{n,j} \ :\ 1 \leq j \leq n,\ n \in \mathbb{N}\},$$

where $x_{n,j}$ are the zeros of $P_n(x)$. Therefore, setting

$$X_1 := Z_1' \equiv \{\text{accumulation points of } Z_1\},$$

$$X_2 := \{x \in Z_1 \ :\ P_n(x) = 0 \text{ for infinitely many } n\},$$

then $\sigma(\psi) \subseteq X_1 \cup X_2$. Moreover, the following holds:

$$\sigma(\psi) \subseteq X_1 \cup X_2 \subseteq [\xi, \eta] = \mathrm{co}(\sigma(\psi)), \tag{5.18}$$

where $\operatorname{co}(\sigma(\psi))$ is the convex hull of the set $\sigma(\psi)$, i.e., it is the smallest closed interval which contains $\sigma(\psi)$, and, as usual, $[\xi,\eta]$ is the true interval of orthogonality of $\{P_n\}_{n\geq 0}$. Indeed, the second inclusion in (5.18) is an immediate consequence of the definitions of the involved sets, and the last equality holds since the interval $\operatorname{co}(\sigma(\psi))$ is a supporting set for $\mathbf{u}$. To prove this, set $\operatorname{co}(\sigma(\psi)) = [a,b]$, and let $p(x)$ be a real polynomial which does not vanish identically on $[a,b]$ and it is non-negative there. Then, we have

$$\langle \mathbf{u}, p\rangle = \int_{-\infty}^{+\infty} p(x)\,\mathrm{d}\psi(x) = \int_a^b p(x)\,\mathrm{d}\psi(x).$$

Since $\sigma(\psi)$ is an infinite set, there is $x_0 \in \sigma(\psi)$ such that $p(x_0) \neq 0$, and so, since $p(x) \geq 0$ for each $x \in [a,b]$, by continuity we have $p(x) > 0$ for each x on a neighborhood $(x_0-\delta, x_0+\delta)$ of the point x_0 –choosing $\delta > 0$ so that $(x_0-\delta, x_0+\delta) \subseteq [a,b]$–, hence, using the Mean Value Theorem for the Riemann-Stieltjes integral, we may write

$$\langle \mathbf{u}, p\rangle = \int_a^b p(x)\,\mathrm{d}\psi(x) \geq \int_{x_0-\delta}^{x_0+\delta} p(x)\,\mathrm{d}\psi(x) = p(\xi_0)\left(\psi(x_0+\delta) - \psi(x_0-\delta)\right),$$

for some $\xi_0 \in (x_0-\delta, x_0+\delta)$. Henceforth, since $p(\xi_0) > 0$ and $\psi(x_0+\delta) - \psi(x_0-\delta) > 0$ ($x_0 \in \sigma(\psi)$), we conclude that $\langle \mathbf{u}, p\rangle > 0$, so that $[a,b] = \operatorname{co}(\sigma(\psi))$ is indeed an interval which is a supporting set for $\mathbf{u}$. Therefore, by Theorem 3.2, the (closed) interval $\operatorname{co}(\sigma(\psi))$ contains the zeros of each P_n, $n \in \mathbb{N}$, hence $[\xi,\eta] \subseteq \operatorname{co}(\sigma(\psi))$. Now, since $\sigma(\psi) \subseteq [\xi,\eta] \subseteq \operatorname{co}(\sigma(\psi))$, we have $[\xi,\eta] = \operatorname{co}(\sigma(\psi))$. Thus (5.18) is proved.

We next show that $X_1 \cup X_2$ is a closed set in $\mathbb{C}$. To prove this fact, we will show that the limit of any convergent sequence of elements in $X_1 \cup X_2$ belongs to this set. Indeed, take arbitrarily a sequence $\{x_n\}_{n\geq 1}$ such that $x_n \in X_1 \cup X_2$ for each $n \in \mathbb{N}$ and $x_n \to x$ in $\mathbb{C}$. We need to prove that $x \in X_1 \cup X_2$. Since $x_n \in X_1 \cup X_2$ for each $n \in \mathbb{N}$, then two situations may occur: $x_n \in X_1$ for infinitely many n, or $x_n \in X_2$ for infinitely many n (or both). In the first situation (passing, if necessary, to a subsequence), since $x_n \to x$ and $X_1 := Z_1'$ is a closed set, we have $x \in X_1$; in the second situation, we have (passing again, if necessary, to a subsequence, and taking into account that $X_2 \subset Z_1$) $x = \lim_{n\to\infty} x_n \in X_2' \subseteq Z_1' = X_1$, X_2' are the accumulation points of X_2. Therefore, in any situation, $x \in X_1 \subseteq X_1 \cup X_2$, which proves that $X_1 \cup X_2$ is closed.

We also need the following result from Complex Analysis, stated here without proof (see e.g., the Remmert book [50], pp. 150–151).

Lemma 5.3 (Vitali Convergence Theorem). *Let G be a domain in $\mathbb{C}$ (i.e., G is a nonempty open connected subset of $\mathbb{C}$), and let $\{f_n\}_{n\geq 1}$ be a* *sequence of analytic functions in G that is locally bounded in G (equivalently, it is bounded on every compact set in G). Suppose that the set*

$$A := \left\{ z \in G : \lim_{n\to\infty} f_n(z) \text{ exists in } \mathbb{C} \right\} \tag{5.19}$$

has at least one accumulation point in G. Then the sequence $\{f_n\}_{n\geq 1}$ converges uniformly on compact subsets of G.

Finally, we are ready to state the Markov theorem.

Theorem 5.2 (Markov Theorem). *Let $\mathbf{u} \in \mathcal{P}'$ be positive definite and let $\{P_n\}_{n\geq 0}$ be the corresponding monic OPS. Let ψ be a natural representative of $\mathbf{u}$, in the sense of Theorem 4.4. Assume further that the spectrum of ψ, $\sigma(\psi)$, is a bounded set. Then*

$$\lim_{n\to+\infty} \frac{u_0 P_{n-1}^{(1)}(z)}{P_n(z)} = \int_{-\infty}^{+\infty} \frac{\mathrm{d}\psi(x)}{z-x}, \quad z \in \mathbb{C}\backslash(X_1 \cup X_2), \tag{5.20}$$

the convergence being uniform on compact subsets of $\mathbb{C}\backslash(X_1 \cup X_2)$.

Proof. Since $\sigma(\psi)$ is bounded, then it follows immediately from (5.18) that the true interval of orthogonality $[\xi, \eta]$ of the sequence $\{P_n\}_{n\geq 0}$ is bounded. According to Lemma 5.1, we may write

$$\frac{u_0 P_{n-1}^{(1)}(z)}{P_n(z)} = \int_{\xi}^{\eta} \frac{\mathrm{d}\psi_n(x)}{z-x}, \quad z \in \mathbb{C}\backslash[\xi, \eta] \tag{5.21}$$

for each $n \in \mathbb{N}$. On the other hand, the representation Theorem 4.4 ensures the existence of a subsequence $\{\psi_{n_j}\}_{j\geq 0}$ which converges on $[\xi, \eta]$ to the given natural representative ψ. It follows from the Helly convergence (Theorem 4.2) that

$$\lim_{j\to+\infty} \frac{u_0 P_{n_j-1}^{(1)}(z)}{P_{n_j}(z)} = \int_{\xi}^{\eta} \frac{\mathrm{d}\psi(x)}{z-x}, \quad z \in \mathbb{C}\backslash[\xi, \eta]. \tag{5.22}$$

Set $M := \max\{|\xi|, |\eta|\}$. We will prove that

$$\lim_{n\to+\infty} \frac{u_0 P_{n-1}^{(1)}(z)}{P_n(z)} = \int_{-\infty}^{+\infty} \frac{\mathrm{d}\psi(x)}{z-x}, \quad |z| > M, \tag{5.23}$$

the convergence being uniform on each set $\{z \in \mathbb{C} : |z| \geq M'\}$ such that

$M' > M$. We start by noticing that, by (5.11), with $k = 0$,

$$\frac{u_0 P_n^{(1)}(z)}{P_{n+1}(z)} - \frac{u_0 P_{n-1}^{(1)}(z)}{P_n(z)} = \frac{C_n}{P_{n+1}(z)P_n(z)}, \quad |z| > M, \tag{5.24}$$

for each $n \in \mathbb{N}_0$, where $C_n := u_0 \prod_{j=1}^{n} \gamma_j$. Since $P_{n+1}(z)P_n(z)$ is a polynomial of degree $2n+1$ with real and simple zeros, then by developing the right-hand side of (5.24) in a Laurent series on the (open) annulus $|z| > M$ (taking into account that

$$\frac{1}{z-x} = \sum_{j \geq 0} x^j / z^{j+1}, \quad |z| > |x|),$$

we see that the Laurent series development of the left-hand side of (5.24) takes the form

$$\frac{u_0 P_n^{(1)}(z)}{P_{n+1}(z)} - \frac{u_0 P_{n-1}^{(1)}(z)}{P_n(z)} = \frac{c_{2n+1}}{z^{2n+1}} + \frac{c_{2n+2}}{z^{2n+2}} + \cdots, \quad |z| > M. \tag{5.25}$$

By repeatedly applying (5.25) we deduce

$$\frac{u_0 P_{m-1}^{(1)}(z)}{P_m(z)} - \frac{u_0 P_{n-1}^{(1)}(z)}{P_n(z)} = \sum_{j=2n}^{\infty} \frac{c_j^{(m,n)}}{z^{j+1}}, \quad |z| > M, \quad m \geq n, \ m, n \in \mathbb{N}_0, \tag{5.26}$$

where the $c_j^{(m,n)}$'s are complex numbers (in fact, we will see that they are real numbers). Next, for each $n \in \mathbb{N}_0$ consider the Laurent series development

$$\frac{u_0 P_{n-1}^{(1)}(z)}{P_n(z)} = \sum_{j=0}^{\infty} \frac{c_j^{(n)}}{z^{j+1}}, \quad |z| > M. \tag{5.27}$$

By comparing (5.26) and (5.27), we deduce

$$c_j^{(m,n)} = c_j^{(m)} - c_j^{(n)}, \quad j \geq 2n, \quad m \geq n, \ m, n \in \mathbb{N}_0. \tag{5.28}$$

Moreover, since

$$\frac{1}{z-x} = \sum_{j \geq 0} \frac{x^j}{z^{j+1}}, \quad |z| > |x|,$$

(5.21) yields

$$\frac{u_0 P_{n-1}^{(1)}(z)}{P_n(z)} = \sum_{j=0}^{\infty} \frac{1}{z^{j+1}} \int_{\xi}^{\eta} x^j \, \mathrm{d}\psi_n(x), \quad |z| > M.$$

Thus, comparing with (5.27), and taking into account the uniqueness of the coefficients of a Laurent development, we obtain

$$c_j^{(n)} = \int_{\xi}^{\eta} x^j \, \mathrm{d}\psi_n(x), \quad j, n \in \mathbb{N}_0.$$

Therefore, for $m \geq n$ and $j \geq 2n$, we deduce

$$\left|c_j^{(m,n)}\right| = \left|c_j^{(m)} - c_j^{(n)}\right| = \left|\int_\xi^\eta x^j \, \mathrm{d}\psi_m(x) - \int_\xi^\eta x^j \, \mathrm{d}\psi_n(x)\right| \leq 2u_0 M^j,$$

where the last inequality holds since $[\xi, \eta] \subseteq [-M, M]$ and $\int_\xi^\eta \mathrm{d}\psi_n(x) = u_0$. Thus, from (5.26) we obtain

$$\left|\frac{u_0 P_{m-1}^{(1)}(z)}{P_m(z)} - \frac{u_0 P_{n-1}^{(1)}(z)}{P_n(z)}\right| \leq 2u_0 \sum_{j=2n}^{\infty} \left(\frac{M}{|z|}\right)^{j+1}, \quad |z| > M, \, m \geq n. \tag{5.29}$$

Since the series

$$\sum_{j=0}^{\infty} \left(\frac{M}{|z|}\right)^{j+1}$$

is convergent whenever $|z| > M$, it follows from (5.29) that $\{u_0 P_{n-1}^{(1)}/P_n\}_{n \geq 0}$ is a Cauchy sequence for each z fulfilling $|z| > M$. Thus, since by (5.22) this sequence has a convergent subsequence, it follows that the sequence converges (to the same limit as its subsequence). Hence (5.23) is proved. Note that the convergence in (5.23) is uniform on each set $\{z \in \mathbb{C} : |z| \geq M'\}$ with $M' > M$. In fact, from (5.29) we obtain

$$\left|\frac{u_0 P_{m-1}^{(1)}(z)}{P_m(z)} - \frac{u_0 P_{n-1}^{(1)}(z)}{P_n(z)}\right| \leq 2u_0 \sum_{j=2n}^{\infty} \left(\frac{M}{M'}\right)^{j+1}, \quad |z| \geq M' > M, \, m \geq n. \tag{5.30}$$

Therefore, since, clearly,

$$\forall \epsilon > 0, \, \exists n_0 \in \mathbb{N} \, : \, \forall n \in \mathbb{N}, \, n \geq n_0 \, \Rightarrow \, 2u_0 \sum_{j=2n}^{\infty} \left(\frac{M}{M'}\right)^{j+1} < \epsilon, \tag{5.31}$$

then $\{u_0 P_{n-1}^{(1)}/P_n\}_{n \geq 0}$ is a (uniformly) Cauchy sequence on the set $\{z \in \mathbb{C} : |z| \geq M'\}$, hence it converges uniformly therein, and so we conclude that, indeed, the convergence in (5.23) holds uniformly on this set. (This fact can be proved directly as follows: Fix $\epsilon > 0$. By (5.31) and (5.30), there exists $n_0 \in \mathbb{N}$ such that

$$\forall m, n \in \mathbb{N}, \, \forall \, z \in A_{M'}, \, m \geq n \geq n_0 \, \Rightarrow \, \left|\frac{u_0 P_{m-1}^{(1)}(z)}{P_m(z)} - \frac{u_0 P_{n-1}^{(1)}(z)}{P_n(z)}\right| < \epsilon,$$

where $A_{M'} := \{z \in \mathbb{C} : |z| \geq M'\}$. Keeping $z \in A_{M'}$ and n fixed, and letting $m \to \infty$, and taking into account that we already proved (5.23) pointwise, it follows that

$$\forall \, m \in \mathbb{N}, \, \forall \, z \in A_{M'}, \, m \geq n_0 \, \Rightarrow \, \left|\int_{-\infty}^{+\infty} \frac{\mathrm{d}\psi(x)}{x - z} - \frac{u_0 P_{n-1}^{(1)}(z)}{P_n(z)}\right| \leq \epsilon,$$

hence the convergence in (5.23) holds uniformly on the set $A_{M'}$, for each $M' > M$.)

To complete the proof, we need to show that the convergence in (5.23) is indeed uniform on compact subsets of $\mathbb{C}\backslash(X_1 \cup X_2)$. For each $N \in \mathbb{N}$, define

$$Z_N := \{ x_{n,j} \, : \, 1 \leq j \leq n, \; n \geq N\}.$$

Let K be a compact subset of $\mathbb{C}\backslash(X_1 \cup X_2)$. Then K contains at most finitely many zeros of the polynomials in the sequence $\{P_n\}_{n\geq 0}$ (otherwise, K could contain an infinite subset of points in Z_1, hence—since it is compact—it would contain a point in $Z_1' = X_1$, and so $K \cap X_1 \neq \emptyset$, which contradicts the definition of K), and none of these zeros belong to X_2. Therefore, there exists $N \in \mathbb{N}$ such that $K \cap (X_1 \cup Z_N) = \emptyset$. Next, set

$$\delta := \text{dist}\,(K, X_1 \cup Z_N) = \inf\{|z - x| \, : \, z \in K, \; x \in X_1 \cup Z_N\}. \qquad (5.32)$$

Since K is compact and $X_1 \cup Z_N$ is closed (this fact can be proved in the same way as we did above to prove that $X_1 \cup X_2$ is closed), and $K \cap (X_1 \cup Z_N) = \emptyset$, then $\delta > 0$. Therefore, using (5.15), we deduce, for each $n \geq N$,

$$\left| \frac{u_0 P_{n-1}^{(1)}(z)}{P_n(z)} \right| \leq \sum_{j=1}^{n} \frac{A_{n,j}}{|z - x_{n,j}|} \leq \frac{1}{\delta} \sum_{j=1}^{n} A_{n,j} = \frac{u_0}{\delta}, \qquad z \in K.$$

(The last inequality follows from (5.32), taking into account that $x_{n,j} \in Z_N$ if $n \geq N$.) Therefore, setting

$$f_n(z) := \frac{P_{n+N-1}^{(1)}(z)}{P_{n+N}(z)}$$

we see that the sequence of functions $\{f_n\}_{n\geq 0}$ is bounded in K. Thus, the claimed result follows from (5.23) and the Vitali convergence theorem (Lemma 5.3), taking therein $G := \mathbb{C}\backslash(X_1 \cup X_2)$ and noting that the set A defined by (5.19) contains $\{z \in \mathbb{C} : |z| > M\}$. □

The statement and proof of the Markov theorem are based on [8, 13, 14, 3]. Markov proved Theorem 5.2 for absolutely continuous measures μ supported on a bounded interval: $\text{supp}(\mu) = [a, b]$. Under such conditions, by (5.18), $X_1 \cup X_2 = [a, b]$. The result remains true for unbounded intervals, provided that the underlying moment problem is determined (see [6]). Different proofs of Markov's theorem, based on the notion of weak convergence of measures, appear in [46, 6].

Remark 5.3. *The set* $X_1 \cup X_2$ *in Theorem 5.2 cannot be replaced by* $\sigma(\psi)$. *For instance, consider the sequence of monic polynomials* $\{P_n\}_{n\geq 0}$ *defined by*

$$P_{2n+1}(x) := 2^n x U_n\left(\frac{x^2-5}{4}\right),$$

$$P_{2n}(x) := 2^n\left(U_n\left(\frac{x^2-5}{4}\right) + 2U_{n-1}\left(\frac{x^2-5}{4}\right)\right).$$

It can be shown that $\{P_n\}_{n\geq 0}$ *is the monic OPS with respect to the measure*

$$\mathrm{d}\mu(x) := \frac{\chi_E(x)}{|x|}\sqrt{1-\left(\frac{x^2-5}{4}\right)^2}\,\mathrm{d}x,$$

where $E := [-3,-1]\cup[1,3]$. *Clearly,* $0 \in X_2 \subseteq X_1 \cup X_2$ *and* $0 \notin E = \sigma(\psi)$. *Nevertheless, the ratio* $P_{n-1}^{(1)}(z)/P_n(z)$ *is not well defined at* $z=0$ *if* n *is odd, hence the sequence* $\left\{P_{n-1}^{(1)}(z)/P_n(z)\right\}_{n\geq 0}$ *has no limit as* $n \to +\infty$ *at* $z=0$.

Exercises

1. Prove that the Stieltjes transform F introduced in Definition 5.1 is an analytic function on $\mathbb{C}\setminus \operatorname{supp}(\mu)$.
2. Let
$$\mathrm{d}\mu(x) := \frac{\chi_{(-1,1)}(x)}{\pi\sqrt{1-x^2}}\,\mathrm{d}x$$
(so that μ is the orthogonality measure for the Chebyshev polynomials of the first kind, $\{T_n\}_{n\geq 0}$). Show that the Stieltjes transform of μ is
$$F(z) = \frac{1}{\sqrt{z^2-1}}, \quad z \in \mathbb{C}\setminus[-1,1],$$
where the branch of the complex square root is chosen so that $\sqrt{z^2-1}$ is an analytic function on $\mathbb{C}\setminus[-1,1]$ and $\sqrt{z^2-1} > 0$ if $z > 1$.
3. Show that the Stieltjes transform of the orthogonality measure
$$\mathrm{d}\mu(x) := \frac{\chi_{(-1,1)}(x)}{\pi}\sqrt{1-x^2}\,\mathrm{d}x$$

(for the Chebyshev polynomials of the second kind, $\{U_n\}_{n\geq 0}$) is

$$F(z) = 2\left(z - \sqrt{z^2-1}\right), \quad z \in \mathbb{C}\setminus[-1,1],$$

where the branch of the complex square root is chosen as in Exercise 2.

4. Prove the relations (5.8) and (5.10).
5. Let $\mathrm{d}\lambda(x) := \chi_{(-1,1)}\mathrm{d}x$ be the orthogonality measure for the Legendre polynomials (so that it is the Lebesgue measure on $[-1,1]$). Show that the associated Legendre polynomials of the first kind are orthogonal with respect to the measure

$$\mathrm{d}\lambda^{(1)}(x) := \frac{2\chi_{(-1,1)}(x)}{\pi^2 + \ln^2 \dfrac{1+x}{1-x}}\,\mathrm{d}x.$$

(Hint. Denote by F and $F^{(1)}$ the Stieltjes transforms of the orthogonality measures for the Legendre polynomials and their associated polynomials of the first kind, respectively.) We may start by showing that

$$F(z) = \mathrm{Log}\left(\frac{z+1}{z-1}\right), \quad z \in \mathbb{C}\setminus[-1,1].$$

Here we took the principal branch of the logarithm so that F is an analytic function on $\mathbb{C}\setminus[-1,1]$. Hence, setting $z = x - i\epsilon$, with $x \in \mathbb{R}$ and $\epsilon > 0$, we deduce

$$F(x-i\epsilon) = \ln\sqrt{\frac{(1+x)^2+\epsilon^2}{(1-x)^2+\epsilon^2}}$$

$$+\, 2i\arctan\left(\sqrt{\left(\frac{x^2-1+\epsilon^2}{2\epsilon}\right)^2 + 1} - \frac{x^2-1+\epsilon^2}{2\epsilon}\right).$$

Next, using the relation $F^{(1)}(z) = z - \beta_0 - u_0/F(z)$, $z \in \mathbb{C}\setminus[-1,1]$ (as usual, u_0 is the moment of order zero for the measure $\mathrm{d}\lambda$, and β_0 is the first β–parameter appearing in the three-term recurrence relation for the monic Legendre polynomials–so that, indeed, we compute $u_0 = 2$ and $\beta_0 = 0$), the orthogonality measure $\mathrm{d}\lambda^{(1)}$ can be easily computed using the Perron-Stieltjes inversion formula, noticing that, writing $\mathrm{d}\lambda^{(1)}(x) = w^{(1)}(x)\mathrm{d}x$, then

$$w^{(1)}(x) = \frac{1}{\pi}\lim_{\epsilon\to 0^+}\Im\left(F^{(1)}(x-i\epsilon)\right), \quad -1 < x < 1.)$$

6. Let $\{U_n\}_{n\geq 0}$ be the Chebyshev OPS of the second kind, which is orthogonal with respect to the positive definite functional $\mathbf{u} \in \mathcal{P}'$ defined by

$$\langle \mathbf{u}, p\rangle := \frac{2}{\pi}\int_{-1}^{1} p(x)\sqrt{1-x^2}\,\mathrm{d}x, \quad p \in \mathcal{P}.$$

Let $\{P_n\}_{n\geq 0}$ be a sequence of polynomials defined by

$$P_{2n+1}(x) := 2^n x U_n\left(\frac{x^2-5}{4}\right),$$

$$P_{2n}(x) := 2^n U_n\left(\frac{x^2-5}{4}\right) + 2^{n-1}U_{n-1}\left(\frac{x^2-5}{4}\right).$$

(a) Prove that $\{P_n\}_{n\geq 0}$ is a monic OPS with respect to a positive definite functional, by showing that it fulfills a three-term recurrence relation

$$P_{-1}(x) = 0, \quad P_0(x) = 1, \quad P_{n+1}(x) = xP_n(x) - \gamma_n P_{n-1}(x),$$

where $\gamma_{2n} = 1$ and $\gamma_{2n+1} = 4$.

(b) Prove that the spectral measure for $\{P_n\}_{n\geq 0}$ (appearing in the spectral theorem for orthogonal polynomials) has distribution function ψ given by

$$\psi(x) := \int_{-\infty}^{x} w(t)\,\mathrm{d}t, \quad x \in \mathbb{R},$$

where

$$\omega(t) := \begin{cases} \dfrac{1}{|t|}\sqrt{1-\left(\dfrac{t^2-5}{4}\right)^2}, & t \in E, \\ 0, & t \notin E, \end{cases}$$

being $E := [-3,-1] \cup [1,3]$. Is the spectral measure unique? Why?

(c) Use the monic OPS $\{P_n\}_{n\geq 0}$ to show that the set $X_1 \cup X_2$ in the statement of Markov's theorem cannot be replaced by $\sigma(\psi)$.

7. Let $\{P_n\}_{n\geq 0}$ be a monic OPS with respect to a positive Borel measure μ. Denote the zeros of P_n by $x_{n,1}, x_{n,2}, \ldots, x_{n,n}$, in increasing order. Let $\{P_n^{(1)}\}_{n\geq 0}$ be the sequence of numerator polynomials, which is a monic OPS with respect to a positive Borel measure $\mu^{(1)}$, and so $P_n^{(1)}$ has n real and simple zeros for each $n \in \mathbb{N}$. Denoting these zeros by $x_{n,1}^{(1)}, x_{n,2}^{(1)}, \ldots, x_{n,n}^{(1)}$, in increasing order, prove the following interlacing property:

$$x_{n+1,j} < x_{n,j}^{(1)} < x_{n+1,j+1}, \quad j = 1, 2, \ldots, n.$$

Conclude that $\mathrm{co}\left(\mathrm{supp}(\mu^{(1)})\right) \subseteq \mathrm{co}\left(\mathrm{supp}(\mu)\right)$.

8. Suppose that $\mathbf{u} \in \mathcal{P}'$ is regular, normalized so that $u_0 := \langle \mathbf{u}, 1 \rangle = 1$, and let $\{P_n\}_{n\geq 0}$ be the monic OPS with respect to $\mathbf{u}$. Let $\lambda \in \mathbb{C} \setminus \{0\}$ and $c \in \mathbb{C}$, and set

$$\mathbf{u}^{\lambda,c} := \delta_c + \lambda(x-c)^{-1}\mathbf{u}.$$

(a) Prove that $\mathbf{u}^{\lambda,c}$ is regular if and only if $P_n(c) + \lambda P_{n-1}^{(1)}(c) \neq 0$ for all $n \in \mathbb{N}$. Moreover, under these conditions, setting

$$a_0 := 0, \quad a_n \equiv a_n^{\lambda,c} := -\frac{P_n(c) + \lambda P_{n-1}^{(1)}(c)}{P_{n-1}(c) + \lambda P_{n-2}^{(1)}(c)} \quad \text{if } n \geq 1,$$

the monic OPS $\{P_n^{\lambda,c}\}_{n\geq 0}$ with respect to $\mathbf{u}^{\lambda,c}$ is given by

$$P_n^{\lambda,c}(x) = P_n(x) + a_n\, P_{n-1}(x),$$

and $\{P_n^{\lambda,c}\}_{n\geq 0}$ fulfills the three-term recurrence relation

$$P_{n+1}^{\lambda,c}(x) = \left(x - \beta_n^{\lambda,c}\right) P_n^{\lambda,c}(x) - \gamma_n^{\lambda,c} P_{n-1}^{\lambda,c}(x), \quad n \geq 0,$$

where $\beta_n^{\lambda,c} := \beta_n + a_n - a_{n+1}$ $(n \geq 0)$, $\gamma_1^{\lambda,c} := \lambda a_1$, and $\gamma_n^{\lambda,c} :=$

$\gamma_{n-1}a_n/a_{n-1}$ ($n \geq 2$), being $\{\beta_n\}_{n\geq 0}$ and $\{\gamma_n\}_{n\geq 1}$ the sequences of parameters appearing in the three-term recurrence relation for $\{P_n\}_{n\geq 0}$, so that $P_{n+1}(x) = (x - \beta_n)P_n(x) - \gamma_n P_{n-1}(x)$, $n \geq 0$, with $\beta_n \in \mathbb{C}$ and $\gamma_n \in \mathbb{C} \setminus \{0\}$ for each n.

(b) Suppose that the $\beta-$parameters vanish in the three-term recurrence relation for $\{P_n\}_{n\geq 0}$ (i.e., $\beta_n = 0$ for each $n \geq 0$). Show that $\mathbf{u}^{\lambda,0} := \delta + \lambda x^{-1}\mathbf{u}$ is regular and the corresponding parameters $a_n \equiv a_n^{\lambda,0}$ defined in (a) are given by

$$a_{2n} = -\frac{1}{\lambda}\frac{P_{2n}(0)}{P_{2n-2}^{(1)}(0)} = \frac{1}{\lambda}\prod_{j=0}^{n-1}\frac{\gamma_{2j+1}}{\gamma_{2j}},$$

$$a_{2n-1} = -\lambda\frac{P_{2n-2}^{(1)}(0)}{P_{2n-2}(0)} = -\lambda\prod_{j=1}^{n-1}\frac{\gamma_{2j}}{\gamma_{2j-1}}$$

for each $n \geq 1$ (with the conventions $\gamma_0 := 1$ and empty product equals 1).

9. (Orthogonal polynomials on the semi-circle) Let $\mathbf{v} \in \mathcal{P}'$ be defined by

$$\langle \mathbf{v}, p\rangle := \frac{1}{\pi}\int_0^\pi p\left(e^{i\theta}\right)\,\mathrm{d}\theta, \quad p \in \mathcal{P}.$$

(a) Show that

$$\mathbf{v} = \delta - \frac{2}{\pi i}x^{-1}\mathbf{u},$$

where $\mathbf{u} \in \mathcal{P}'$ is the (positive definite) Legendre functional normalized so that

$$\langle \mathbf{u}, p\rangle := \frac{1}{2}\int_{-1}^{1} p(x)\,\mathrm{d}x, \quad p \in \mathcal{P}.$$

(Hint. Note that

$$\int_\Gamma \frac{p(z)}{z}\,\mathrm{d}z = 0$$

for each $p \in \mathcal{P}$, where Γ is the closed path on $\mathbb{C}$ defined by $\Gamma := \Gamma_1 + \ell_\epsilon^- + \Gamma_\epsilon + \ell_\epsilon^+$, Γ_1 and Γ_ϵ being semicircles on the upper semi-plane of radius 1 and ϵ, respectively, with $0 < \epsilon < 1$, Γ_1 starting at the point $z = 1$ and ending at $z = -1$, and Γ_ϵ starting at $z = -\epsilon$ and ending at $z = \epsilon$, and ℓ_ϵ^- and ℓ_ϵ^+ are segments on the real line, joining the points $z = -1$ to $z = -\epsilon$, and $z = \epsilon$ to $z = 1$, respectively. Consider the integrals along each of the paths Γ_1, ℓ_ϵ^-, Γ_ϵ, and ℓ_ϵ^+, and then take the limit as $\epsilon \to 0^+$.)

(b) Prove that $\mathbf{v}$ is regular and the monic OPS $\{Q_n\}_{n\geq 0}$ with respect to $\mathbf{v}$ is given by

$$Q_n(x) = P_n(x) - \frac{2i}{2n-1}\left(\frac{\Gamma\left(\frac{n+1}{2}\right)}{\Gamma\left(\frac{n}{2}\right)}\right)^2 P_{n-1}(x), \quad n \in \mathbb{N},$$

where $\{P_n\}_{n\geq 0}$ is the (Legendre) monic OPS with respect to $\mathbf{u}$. (Hint. Use Exercise 8)

Chapter 6

Orthogonal Polynomials and Dual Basis

In this chapter we present several properties of the dual basis associated with an orthogonal polynomial sequence. We also introduce the translation and homothetic operators in $\mathcal{P}$ and $\mathcal{P}^*$, which appear as useful tools in the study of several classes of orthogonal polynomials, including the classical families.

6.1 Orthogonal Polynomials and Dual Basis

We begin by establishing some connections between a regular functional and the dual basis associated with the corresponding monic OPS.

Theorem 6.1. *Let $\mathbf{u} \in \mathcal{P}'$ be regular and let $\{P_n\}_{n\geq 0}$ be the corresponding monic OPS with associated dual basis $\{\mathbf{a}_n\}_{n\geq 0}$. The following hold:*

(i) *For each $n \in \mathbb{N}_0$, $\mathbf{a}_n$ is explicitly given by*

$$\mathbf{a}_n = \frac{P_n}{\langle \mathbf{u}, P_n^2 \rangle}\,\mathbf{u}.$$

As a consequence, $\{P_n\}_{n\geq 0}$ is a monic OPS with respect to $\mathbf{a}_0$, where

$$\mathbf{u} = u_0\,\mathbf{a}_0,$$

with $u_0 := \langle \mathbf{u}, 1 \rangle$.

(ii) *Let $\mathbf{v} \in \mathcal{P}'$ and $N \in \mathbb{N}_0$ such that*

$$\langle \mathbf{v}, P_n \rangle = 0 \text{ if } n \geq N+1.$$

Then,

$$\mathbf{v} = \sum_{j=0}^{N} \langle \mathbf{v}, P_j \rangle\,\mathbf{a}_j = \phi\,\mathbf{u}, \quad \phi(x) := \sum_{j=0}^{N} \frac{\langle \mathbf{v}, P_j \rangle}{\langle \mathbf{u}, P_j^2 \rangle} P_j(x).$$

Further, $\deg \phi \leq N$, and $\deg \phi = N$ if and only if $\langle \mathbf{v}, P_N \rangle \neq 0$.

DOI: 10.1201/9781003432067-6

(iii) *If* $\{P_n\}_{n\geq 0}$ *satisfies*

$$xP_n(x) = P_{n+1}(x) + \beta_n P_n(x) + \gamma_n P_{n-1}(x), \quad n \in \mathbb{N}_0,$$

with $P_{-1}(x) = 0$, $P_0(x) = 1$, $\beta_n \in \mathbb{C}$, *and* $\gamma_n \in \mathbb{C}\backslash\{0\}$, *then* $\{\mathbf{a}_n\}_{n\geq 0}$ *satisfies*

$$x\,\mathbf{a}_n = \mathbf{a}_{n-1} + \beta_n\,\mathbf{a}_n + \gamma_{n+1}\,\mathbf{a}_{n+1},$$

with initial conditions $\mathbf{a}_{-1} = \mathbf{0}$ *and* $\mathbf{a}_0 = u_0^{-1}\,\mathbf{u}$.

Proof. By Theorem 1.1, for each $n \in \mathbb{N}_0$ we may write

$$P_n\mathbf{u} = \sum_{j\geq 0}\langle P_n\mathbf{u}, P_j\rangle\mathbf{a}_j = \sum_{j\geq 0}\langle \mathbf{u}, P_nP_j\rangle\mathbf{a}_j = \langle \mathbf{u}, P_n^2\rangle\mathbf{a}_n,$$

and (i) is proved. The statement (ii) follows immediately from (i) using again Theorem 1.1. Finally, for all $n \in \mathbb{N}$ and $j \in \mathbb{N}_0$ we have

$$\begin{aligned} x\,\mathbf{a}_n &= \frac{xP_n}{\langle \mathbf{u}, P_n^2\rangle}\,\mathbf{u} = \frac{P_{n+1} + \beta_nP_n + \gamma_nP_{n-1}}{\langle \mathbf{u}, P_n^2\rangle}\,\mathbf{u} \\ &= \frac{\langle \mathbf{u}, P_{n+1}^2\rangle}{\langle \mathbf{u}, P_n^2\rangle}\frac{P_{n+1}}{\langle \mathbf{u}, P_{n+1}^2\rangle}\mathbf{u} + \beta_n\frac{P_n}{\langle \mathbf{u}, P_n^2\rangle}\,\mathbf{u} + \gamma_n\frac{\langle \mathbf{u}, P_{n-1}^2\rangle}{\langle \mathbf{u}, P_n^2\rangle}\frac{P_{n-1}}{\langle \mathbf{u}, P_{n-1}^2\rangle}\mathbf{u} \\ &= \gamma_{n+1}\,\mathbf{a}_{n+1} + \beta_n\,\mathbf{a}_n + \mathbf{a}_{n-1}, \end{aligned}$$

where we have used the relation $\gamma_i = \langle \mathbf{u}, P_i^2\rangle/\langle \mathbf{u}, P_{i-1}^2\rangle$, $i \in \mathbb{N}$, see Corollary 2.4. □

Corollary 6.1. *Let* $\{P_n\}_{n\geq 0}$ *be a monic OPS and let* $\mathbf{v} \in \mathcal{P}'$. *Then* $\{P_n\}_{n\geq 0}$ *is a monic OPS with respect to* $\mathbf{v}$ *if and only if*

$$\langle \mathbf{v}, 1\rangle \neq 0, \qquad \langle \mathbf{v}, P_n\rangle = 0, \quad n \in \mathbb{N}. \tag{6.1}$$

Proof. Clearly, if $\{P_n\}_{n\geq 0}$ is a monic OPS with respect to $\mathbf{v}$ then (6.1) holds. Conversely, if (6.1) holds, then by (ii) in Theorem 6.1, we get

$$\mathbf{v} = \langle \mathbf{v}, 1\rangle\,\mathbf{a}_0 = \frac{\langle \mathbf{v}, 1\rangle}{\langle \mathbf{u}, 1\rangle}\,\mathbf{u},$$

where $\mathbf{a}_0$ is the first element of the dual basis associated with $\{P_n\}_{n\geq 0}$, and $\mathbf{u}$ is the regular functional with respect to which $\{P_n\}_{n\geq 0}$ is an OPS. Since, by hypothesis, $\langle \mathbf{v}, 1\rangle \neq 0$, it follows that $\{P_n\}_{n\geq 0}$ is a monic OPS with respect to $\mathbf{v}$. □

6.2 The Translation and Homothetic Operators

Definition 6.1 (Translation Operators). *Let $b \in \mathbb{C}$.*

(i) *The translator operator on $\mathcal{P}$ is $\tau_b : \mathcal{P} \to \mathcal{P}$ $(p \mapsto \tau_b p)$ defined by*

$$\tau_b p(x) := p(x-b), \quad p \in \mathcal{P}; \tag{6.2}$$

(ii) *The translator operator on $\mathcal{P}'$ is $\boldsymbol{\tau}_b := \tau'_{-b}$, i.e., $\boldsymbol{\tau}_b : \mathcal{P}' \to \mathcal{P}'$ is the dual operator of τ_{-b}, so that*

$$\langle \boldsymbol{\tau}_b \mathbf{u}, p \rangle := \langle \mathbf{u}, \tau_{-b} p \rangle = \langle \mathbf{u}, p(x+b) \rangle, \quad \mathbf{u} \in \mathcal{P}', \quad p \in \mathcal{P}. \tag{6.3}$$

Notice that the moments of the functional $\boldsymbol{\tau}_b \mathbf{u}$ are

$$(\boldsymbol{\tau}_b \mathbf{u})_n = \sum_{j=0}^{n} \binom{n}{j} b^{n-j} u_j = \sum_{i+j=n} \binom{n}{i} b^i u_j, \quad n \in \mathbb{N}_0. \tag{6.4}$$

Indeed, for each $n \in \mathbb{N}_0$,

$$(\boldsymbol{\tau}_b \mathbf{u})_n := \langle \boldsymbol{\tau}_b \mathbf{u}, x^n \rangle = \langle \mathbf{u}, \tau_{-b} x^n \rangle = \langle \mathbf{u}, (x+b)^n \rangle = \sum_{j=0}^{n} \binom{n}{j} b^{n-j} \langle \mathbf{u}, x^j \rangle.$$

Definition 6.2 (Homothetic Operators). *Let $a \in \mathbb{C} \setminus \{0\}$.*

(i) *The homothetic operator on $\mathcal{P}$ is $h_a : \mathcal{P} \to \mathcal{P}$ $(p \mapsto h_a p)$ defined by*

$$h_a p(x) := p(ax), \quad p \in \mathcal{P}. \tag{6.5}$$

(ii) *The homothetic operator on $\mathcal{P}'$ is $\mathbf{h}_a := h'_a$, i.e., $\mathbf{h}_a : \mathcal{P}' \to \mathcal{P}'$ is the dual operator of h_a, so that*

$$\langle \mathbf{h}_a \mathbf{u}, p \rangle := \langle \mathbf{u}, h_a p \rangle = \langle \mathbf{u}, p(ax) \rangle, \quad \mathbf{u} \in \mathcal{P}', \quad p \in \mathcal{P}. \tag{6.6}$$

The moments of the functional $\mathbf{h}_a \mathbf{u}$ are

$$(\mathbf{h}_a \mathbf{u})_n = a^n u_n, \quad n \in \mathbb{N}_0. \tag{6.7}$$

In the next proposition we list some useful properties involving the translation and homothetic operators.

Proposition 6.1. *Let $a \in \mathbb{C} \setminus \{0\}$, $b \in \mathbb{C}$, $\mathbf{u} \in \mathcal{P}'$, and $p \in \mathcal{P}$. Then the following properties hold:*

1. $\tau_0 p = h_1 p = p$,
2. $(\tau_b \circ \tau_{-b})\, p = (\tau_{-b} \circ \tau_b)\, p = p$,
3. $(h_a \circ h_{a^{-1}})\, p = (h_{a^{-1}} \circ h_a)\, p = p$,
4. $(h_a \circ \tau_b)\, p = \left(\tau_{b/a} \circ h_a\right) p$,
5. $(\tau_b \circ h_a)\, p = (h_a \circ \tau_{ab})\, p$,
6. $\boldsymbol{\tau}_0 \mathbf{u} = \mathbf{h}_1 \mathbf{u} = \mathbf{u}$,
7. $(\boldsymbol{\tau}_b \circ \boldsymbol{\tau}_{-b})\, \mathbf{u} = (\boldsymbol{\tau}_{-b} \circ \boldsymbol{\tau}_b)\, \mathbf{u} = \mathbf{u}$,
8. $(\mathbf{h}_a \circ \mathbf{h}_{a^{-1}})\, \mathbf{u} = (\mathbf{h}_{a^{-1}} \circ \mathbf{h}_a)\, \mathbf{u} = \mathbf{u}$,
9. $(\mathbf{h}_a \circ \boldsymbol{\tau}_b)\, \mathbf{u} = (\boldsymbol{\tau}_{ab} \circ \mathbf{h}_a)\, \mathbf{u}$,
10. $(\boldsymbol{\tau}_b \circ \mathbf{h}_a)\, \mathbf{u} = \left(\mathbf{h}_a \circ \boldsymbol{\tau}_{b/a}\right) \mathbf{u}$.

Proof. Properties 1, 2, 3, 6, 7, and 8 follow easily by straightforward computations. Note that

$$(h_a \circ \tau_b)\, p(x) = h_a\left((\tau_b p)(x)\right) = (\tau_b p)(ax) = p(ax - b) = p\left(a\left(x - \frac{b}{a}\right)\right)$$

$$= \tau_{b/a}\left(p(ax)\right) = \tau_{b/a}\left(h_a p(x)\right) = \left(\tau_{b/a} \circ h_a\right) p(x),$$

and Property 4 follows. Replacing b by ab in Property 4 we obtain Property 5. To prove Property 9, note that

$$\langle (\mathbf{h}_a \circ \boldsymbol{\tau}_b)\, \mathbf{u}, p \rangle = \langle \boldsymbol{\tau}_b \mathbf{u}, h_a p \rangle = \langle \mathbf{u}, (\tau_{-b} \circ h_a)\, p \rangle = \langle \mathbf{u}, (h_a \circ \tau_{-ab})\, p \rangle$$

$$= \langle \mathbf{h}_a \mathbf{u}, \tau_{-ab} p \rangle = \langle (\boldsymbol{\tau}_{ab} \circ \mathbf{h}_a)\, \mathbf{u}, p \rangle,$$

where in the third equality we have used Property 5. Finally, replacing b by b/a in Property 9 we obtain Property 10. □

Properties 2 and 3 show that the operators τ_b and h_a are invertible in $\mathcal{P}$:

$$\tau_b^{-1} p = \tau_{-b} p, \quad h_a^{-1} p = h_{a^{-1}} p, \quad a \in \mathbb{C} \setminus \{0\},\ b \in \mathbb{C},\ p \in \mathcal{P}. \tag{6.8}$$

Similarly, Properties 6 and 7 show that $\boldsymbol{\tau}_b$ and $\mathbf{h}_a$ are invertible in $\mathcal{P}'$:

$$\boldsymbol{\tau}_b^{-1} \mathbf{u} = \boldsymbol{\tau}_{-b} \mathbf{u}, \quad \mathbf{h}_a^{-1} \mathbf{u} = \mathbf{h}_{a^{-1}} \mathbf{u}, \quad a \in \mathbb{C} \setminus \{0\},\ b \in \mathbb{C},\ \mathbf{u} \in \mathcal{P}'. \tag{6.9}$$

As a consequence, we also deduce

$$(h_a \circ \tau_b)^{-1} = \tau_{-b} \circ h_{a^{-1}}, \qquad (\tau_b \circ h_a)^{-1} = h_{a^{-1}}, \circ \tau_{-b},$$

$$(\mathbf{h}_a \circ \boldsymbol{\tau}_b)^{-1} = \boldsymbol{\tau}_{-b} \circ \mathbf{h}_{a^{-1}}, \qquad (\boldsymbol{\tau}_b \circ \mathbf{h}_a)^{-1} = \mathbf{h}_{a^{-1}} \circ \boldsymbol{\tau}_{-b}.$$

Finally, we point out the following property that one should keep in mind (it follows immediately from the proof of Property 4 replacing b by $-b$):

$$(h_a \circ \tau_{-b})\, p(x) = p(ax+b), \quad a \in \mathbb{C}\setminus\{0\},\, b \in \mathbb{C},\, p \in \mathcal{P}. \tag{6.10}$$

The next proposition is of fundamental importance for a rigorous treatment of the classification of classical OP.

Theorem 6.2. *Define a binary relation on $\mathcal{P}'$ as follows: for every $\mathbf{u}, \mathbf{v} \in \mathcal{P}'$,*

$$\mathbf{u} \sim \mathbf{v} \quad \text{iff} \quad \exists a \in \mathbb{C}\setminus\{0\},\, \exists b \in \mathbb{C} \,:\, \mathbf{v} = (\mathbf{h}_{a^{-1}} \circ \boldsymbol{\tau}_{-b})\,\mathbf{u}. \tag{6.11}$$

Then, $\sim$ is an equivalent relation on $\mathcal{P}'$.

Proof. Since

$$\mathbf{u} = (\mathbf{h}_1 \circ \boldsymbol{\tau}_0)\,\mathbf{u},$$

then $\mathbf{u} \sim \mathbf{u}$, so that the binary relation $\sim$ is reflexive. To prove that it is symmetric, assume that $\mathbf{u} \sim \mathbf{v}$. Then (6.11) holds. Hence

$$\mathbf{u} = (\mathbf{h}_{a^{-1}} \circ \boldsymbol{\tau}_{-b})^{-1}\,\mathbf{v} = (\boldsymbol{\tau}_b \circ \mathbf{h}_a)\,\mathbf{v} = \left(\mathbf{h}_a \circ \boldsymbol{\tau}_{b/a}\right)\mathbf{v} = (\mathbf{h}_{c^{-1}} \circ \boldsymbol{\tau}_{-d})\,\mathbf{v},$$

where $c := a^{-1} \in \mathbb{C}\setminus\{0\}$ and $d := -b/a \in \mathbb{C}$. (Notice also that the third equality follows from Property 10 in Proposition 6.1.) Thus, $\mathbf{v} \sim \mathbf{u}$. Finally, to prove that $\sim$ is transitive, suppose that $\mathbf{u} \sim \mathbf{v}$ and $\mathbf{v} \sim \mathbf{w} \in \mathcal{P}'$. Then, there exists $a, c \in \mathbb{C}\setminus\{0\}$ and $b, d \in \mathbb{C}$ such that

$$\mathbf{v} = (\mathbf{h}_{a^{-1}} \circ \boldsymbol{\tau}_{-b})\,\mathbf{u}, \quad \mathbf{w} = (\mathbf{h}_{c^{-1}} \circ \boldsymbol{\tau}_{-d})\,\mathbf{v}.$$

As a consequence, we may write

$$\begin{aligned}
\mathbf{w} &= (\mathbf{h}_{c^{-1}} \circ \boldsymbol{\tau}_{-d})\,(\mathbf{h}_{a^{-1}} \circ \boldsymbol{\tau}_{-b})\,\mathbf{u} = (\mathbf{h}_{c^{-1}} \circ (\boldsymbol{\tau}_{-d} \circ \mathbf{h}_{a^{-1}}) \circ \boldsymbol{\tau}_{-b})\,\mathbf{u} \\
&= (\mathbf{h}_{c^{-1}} \circ (\mathbf{h}_{a^{-1}} \circ \boldsymbol{\tau}_{-ad}) \circ \boldsymbol{\tau}_{-b})\,\mathbf{u} = ((\mathbf{h}_{c^{-1}} \circ \mathbf{h}_{a^{-1}}) \circ (\boldsymbol{\tau}_{-ad} \circ \boldsymbol{\tau}_{-b}))\,\mathbf{u} \\
&= (\mathbf{h}_{a^{-1}c^{-1}} \circ \boldsymbol{\tau}_{-ad-b})\,\mathbf{u} = (\mathbf{h}_{\alpha^{-1}} \circ \boldsymbol{\tau}_{-\beta})\,\mathbf{u},
\end{aligned}$$

where $\alpha := ac \in \mathbb{C}\setminus\{0\}$ and $\beta := b + ad \in \mathbb{C}$. Thus $\mathbf{u} \sim \mathbf{w}$, which completes the proof. □

Remark 6.1. *The relation between $\mathbf{u}$ and $\mathbf{v}$ in (6.11) may be expressed as*

$$\langle \mathbf{v}, x^n \rangle = \left\langle \mathbf{u}, \left(\frac{x-b}{a}\right)^n \right\rangle, \quad n \in \mathbb{N}_0. \tag{6.12}$$

Theorem 6.3. *Let $\{P_n\}_{n\geq 0}$ be a simple set in $\mathcal{P}$ and let $\{\mathbf{a}_n\}_{n\geq 0}$ be the associated dual basis. Let $a \in \mathbb{C} \setminus \{0\}$ and $b \in \mathbb{C}$. Define*

$$Q_n := a^{-n} \left(h_a \circ \tau_{-b}\right) P_n. \tag{6.13}$$

Then $\{Q_n\}_{n\geq 0}$ is a simple set in $\mathcal{P}$, and its dual basis, $\{\mathbf{b}_n\}_{n\geq 0}$, is given by

$$\mathbf{b}_n = a^n \left(\mathbf{h}_{a^{-1}} \circ \boldsymbol{\tau}_{-b}\right) \mathbf{a}_n. \tag{6.14}$$

Proof. It is clear that $\{Q_n\}_{n\geq 0}$ is a simple set in $\mathcal{P}$. Moreover, for every $n, k \in \mathbb{N}_0$,

$$\begin{aligned}
\langle \mathbf{b}_n, Q_k \rangle &= a^{n-k} \langle (\mathbf{h}_{a^{-1}} \circ \boldsymbol{\tau}_{-b})\, \mathbf{a}_n, (h_a \circ \tau_{-b})\, P_k \rangle \\
&= a^{n-k} \langle \mathbf{a}_n, (\tau_b \circ h_{a^{-1}})\, (h_a \circ \tau_{-b})\, P_k \rangle \\
&= a^{n-k} \langle \mathbf{a}_n, P_k \rangle = a^{n-k} \delta_{n,k} = \delta_{n,k}.
\end{aligned}$$

Thus $\{\mathbf{b}_n\}_{n\geq 0}$ is the dual basis associated with $\{Q_n\}_{n\geq 0}$. □

Remark 6.2. *By (6.10) we see that the polynomial Q_n in (6.13) is indeed*

$$Q_n(x) := a^{-n} P_n(ax + b), \tag{6.15}$$

so that Q_n is obtained from P_n by an affine change of the variable, being Q_n normalized so that it becomes a monic polynomial whenever P_n is monic.

Theorem 6.4. *Under the hypothesis of Theorem 6.3, assume further that $\{P_n\}_{n\geq 0}$ is a monic OPS with respect to the functional $\mathbf{u} \in \mathcal{P}'$, and let*

$$xP_n(x) = P_{n+1}(x) + \beta_n P_n(x) + \gamma_n P_{n-1}(x), \tag{6.16}$$

be the recurrence relation fulfilled by $\{P_n\}_{n\geq 0}$, with initial conditions $P_{-1}(x) = 0$ and $P_0(x) = 1$, being $\beta_n \in \mathbb{C}$ and $\gamma_n \in \mathbb{C} \setminus \{0\}$. Then, $\{Q_n\}_{n\geq 0}$ is a monic OPS with respect to

$$\mathbf{v} := \left(\mathbf{h}_{a^{-1}} \circ \boldsymbol{\tau}_{-b}\right) \mathbf{u}, \tag{6.17}$$

and the recurrence relation fulfilled by $\{Q_n\}_{n\geq 0}$ is

$$xQ_n(x) = Q_{n+1}(x) + \widehat{\beta}_n Q_n(x) + \widehat{\gamma}_n Q_{n-1}(x), \tag{6.18}$$

with initial conditions $Q_{-1}(x) = 0$ and $Q_0(x) = 1$, where

$$\widehat{\beta}_n := \frac{\beta_n - b}{a}, \quad \widehat{\gamma}_n := \frac{\gamma_n}{a^2}. \tag{6.19}$$

Proof. Changing x into $ax+b$ in (6.16) and then multiplying both sides of the resulting equality by a^{-n-1}, we obtain (6.18). Since $\{Q_n\}_{n\geq 0}$ satisfies (6.18) and $\widehat{\gamma}_n \neq 0$ for each $n \geq 1$, then it is a monic OPS (by Favard's Theorem). By Theorem 6.3, the dual basis associated with $\{Q_n\}_{n\geq 0}$ is given by (6.14). Moreover, by Theorem 6.1–(i), $\{Q_n\}_{n\geq 0}$ is a monic OPS with respect to $\mathbf{b}_0$. Therefore, since

$$\mathbf{b}_0 = (\mathbf{h}_{a^{-1}} \circ \boldsymbol{\tau}_{-b})\,\mathbf{a}_0 = (\mathbf{h}_{a^{-1}} \circ \boldsymbol{\tau}_{-b})\,u_0^{-1}\mathbf{u} = u_0^{-1}\mathbf{v},$$

so that $\mathbf{v} = u_0\mathbf{b}_0$ $(u_0 \neq 0)$, we conclude that $\{Q_n\}_{n\geq 0}$ is a monic OPS with respect to $\mathbf{v}$. □

Exercises

1. Let $\mathbf{u} \in \mathcal{P}'$ be regular and let $\{P_n\}_{n\geq 0}$ be the corresponding monic OPS with associated dual basis and $\{\mathbf{a}_n\}_{n\geq 0}$. Consider the monic OPS $\{P_n^{(k)}\}_{n\geq 0}$ (associated polynomials of order $k \in \mathbb{N}$) and let $\{\mathbf{a}_n^{(k)}\}_{n\geq 0}$ be the corresponding dual basis. Show that
$$x^{-1}\left(\mathbf{a}_n^{(k)}\mathbf{a}_{k-1}\right) = \mathbf{a}_{n+k}.$$
In particular, $\mathbf{a}_n^{(1)} = (x\mathbf{a}_{n+1})\,\mathbf{a}_0^{-1}$, and so $\{P_n^{(1)}\}_{n\geq 0}$ is an OPS with respect to the functional $\mathbf{u}^{(1)}$ given by
$$\mathbf{u}^{(1)} = c\,(xP_1\mathbf{u})\mathbf{u}^{-1} = -cu_0x^2\mathbf{u}^{-1}, \quad c := u_0^{(1)}/\gamma_1 \in \mathbb{C}\setminus\{0\}.$$

Chapter 7

Functional Differential Equation

In this chapter we present necessary and sufficient conditions for the regularity of solutions of the functional equation $D(\phi\mathbf{u}) = \psi\mathbf{u}$, where $\phi \in \mathcal{P}_2$ and $\psi \in \mathcal{P}_1$. Moreover, the functional Rodrigues formula and a closed formula for the recurrence coefficients are presented.

7.1 The Functional Differential Equation

The functional differential equation considered in this chapter has the form

$$D(\phi\mathbf{u}) = \psi\mathbf{u}, \tag{7.1}$$

where $\phi \in \mathcal{P}_2$ and $\psi \in \mathcal{P}_1$, and $\mathbf{u} \in \mathcal{P}'$ is the unknown. Notice that we do not require *a priori* $\mathbf{u}$ to be a regular functional. Define

$$\phi(x) := ax^2 + bx + c, \quad \psi(x) := px + q, \tag{7.2}$$

being $a, b, c, p, q \in \mathbb{C}$. We also define, for each integer or rational number n,

$$\psi_n := \psi + n\phi', \quad d_n := \psi'_{n/2} = na + p, \quad e_n := \psi_n(0) = nb + q. \tag{7.3}$$

Notice that $\psi_n(x) = d_{2n}x + e_n \in \mathcal{P}_1$. Finally, for each $\mathbf{u} \in \mathcal{P}'$ and each $n \in \mathbb{N}_0$, we set

$$\mathbf{u}^{[n]} := \phi^n\mathbf{u}. \tag{7.4}$$

We start with the following elementary result.

Lemma 7.1. *The functional $\mathbf{u} \in \mathcal{P}'$ satisfies the functional differential equation (7.1), with ϕ and ψ given by (7.2), if and only if the corresponding sequence of moments,*

$$u_n := \langle \mathbf{u}, x^n \rangle, \quad n \in \mathbb{N}_0,$$

satisfies the second order linear difference equation

$$d_n u_{n+1} + e_n u_n + n\phi(0)u_{n-1} = 0. \tag{7.5}$$

DOI: 10.1201/9781003432067-7

Moreover, if $\mathbf{u}$ *satisfies* (7.1), *then* $\mathbf{u}^{[n]}$ *satisfies*

$$D\left(\phi\mathbf{u}^{[n]}\right) = \psi_n\mathbf{u}^{[n]}. \tag{7.6}$$

Proof. An easy computation shows that

$$\begin{aligned} D(\phi\mathbf{u}) = \psi\mathbf{u} \quad &\Leftrightarrow \quad \langle D(\phi\mathbf{u}), x^n\rangle = \langle\psi\mathbf{u}, x^n\rangle, \\ &\Leftrightarrow \quad -n\langle\mathbf{u}, \phi x^{n-1}\rangle = \langle\mathbf{u}, \psi x^n\rangle, \\ &\Leftrightarrow \quad (na+p)u_{n+1} + (nb+q)u_n + ncu_{n-1} = 0, \end{aligned}$$

and the first assertion of the lemma is proved. The proof of (7.6) is by induction on n. Since $\mathbf{u}^{[0]} := \mathbf{u}$ and $\psi_0 = \psi$, then (7.6) holds for $n = 0$. Assume that (7.6) holds for a certain $n \in \mathbb{N}$. Hence

$$D\left(\mathbf{u}^{[n+1]}\right) = D\left(\phi\mathbf{u}^{[n]}\right) = \psi_n\mathbf{u}^{[n]}.$$

Thus

$$\begin{aligned} D\left(\phi\mathbf{u}^{[n+1]}\right) &= \phi'\mathbf{u}^{[n+1]} + \phi D\left(\mathbf{u}^{[n+1]}\right) \\ &= \phi'\mathbf{u}^{[n+1]} + \phi\psi_n\mathbf{u}^{[n]} = (\phi' + \psi_n)\,\mathbf{u}^{[n+1]} = \psi_{n+1}\mathbf{u}^{[n+1]}, \end{aligned}$$

and (7.6) is proved. □

Theorem 7.1. *Suppose that* $\mathbf{u} \in \mathcal{P}'$ *satisfies the functional differential equation* (7.1), *with* ϕ *and* ψ *given by* (7.2). *Assume further that*

$$d_n \neq 0, \quad n \in \mathbb{N}_0. \tag{7.7}$$

Then, there exists a simple set of polynomials $\{R_n\}_{n\geq 0}$ *such that*

$$R_n\mathbf{u} = D^n(\phi^n\mathbf{u}). \tag{7.8}$$

Moreover, $\{R_n\}_{n\geq 0}$ *may be chosen so that it satisfies the three-term recurrence relation*

$$R_{n+1}(x) = (\widetilde{\alpha}_n x - \widetilde{\beta}_n)R_n(x) - \widetilde{\gamma}_n R_{n-1}(x), \tag{7.9}$$

with initial conditions $R_{-1}(x) = 0$ and $R_0(x) = 1$, where

$$\widetilde{\alpha}_n := \frac{d_{2n-1}d_{2n}}{d_{n-1}},$$

$$\widetilde{\beta}_n := -\widetilde{\alpha}_n \frac{d_{-2}q + 2\,b\,n\,d_{n-1}}{d_{2n-2}d_{2n}},$$

$$\widetilde{\gamma}_n := -\widetilde{\alpha}_n \frac{nd_{2n-2}}{d_{2n-1}}\,\phi\left(-\frac{e_{n-1}}{d_{2n-2}}\right).$$

Proof. The proof is by induction on n. Set

$$R_0 := 1, \quad R_1 := \psi.$$

It is clear that (7.8) holds for $n = 0$ and $n = 1$. Suppose now that (7.8) holds for the indices n and $n-1$, that is, there exist polynomials R_n and R_{n-1}, with degrees n and $n-1$, respectively, such that

$$R_n\mathbf{u} = D^n\left(\phi^n\mathbf{u}\right), \quad R_{n-1}\mathbf{u} = D^{n-1}\left(\phi^{n-1}\mathbf{u}\right). \tag{7.10}$$

We must show that there is a polynomial R_{n+1} of degree $n+1$ such that

$$R_{n+1}\mathbf{u} = D^{n+1}\left(\phi^{n+1}\mathbf{u}\right). \tag{7.11}$$

By Lemma 7.1, we get

$$\begin{aligned} D^{n+1}\left(\phi^{n+1}\mathbf{u}\right) &= D^n\left(D\left(\phi\mathbf{u}^{[n]}\right)\right) = D^n\left(\psi_n\mathbf{u}^{[n]}\right) = D^n\left(\psi_n\phi^n\mathbf{u}\right) \\ &= \binom{n}{0}\psi_n D^n(\phi^n\mathbf{u}) + \binom{n}{1}\psi_n' D^{n-1}(\phi^n\mathbf{u}) \\ &= \psi_n R_n\mathbf{u} + n\psi_n' D^{n-1}(\phi^n\mathbf{u}), \end{aligned}$$

where in the fourth equality we applied the Leibniz formula (see Chapter 1) and in the last one we used the first relation in the induction hypothesis (7.10). Hence

$$D^{n-1}(\phi^n\mathbf{u}) = \frac{1}{nd_{2n}}\left(D^{n+1}(\phi^{n+1}\mathbf{u}) - \psi_n R_n\mathbf{u}\right). \tag{7.12}$$

Notice that, according to the hypothesis (7.7), $\psi'_{m/2} = d_m \neq 0$ for each $m \in \mathbb{N}_0$. We point out that we have deduced (7.12) using the first relation in (7.10). Therefore, making the change of indices $n \to n-1$ in the above reasoning and using the second relation in (7.10), we obtain

$$D^{n-2}(\phi^{n-1}\mathbf{u}) = \frac{1}{(n-1)d_{2n-2}}(R_n - \psi_{n-1}R_{n-1})\mathbf{u}. \tag{7.13}$$

On the other hand, using again Lemma 7.1, we have

$$D^{n+1}(\phi^{n+1}\mathbf{u}) = D^{n-1}\left(D(\psi_n\phi^n\mathbf{u})\right) = D^{n-1}\left((\psi_n'\phi + \psi_n\psi_{n-1})\phi^{n-1}\mathbf{u}\right)$$

$$= (\psi_n'\phi + \psi_n\psi_{n-1})D^{n-1}(\phi^{n-1}\mathbf{u})$$

$$+ \binom{n-1}{1}(\psi_n'\phi + \psi_n\psi_{n-1})'D^{n-2}(\phi^{n-1}\mathbf{u})$$

$$+ \binom{n-1}{2}(\psi_n'\phi + \psi_n\psi_{n-1})''D^{n-3}(\phi^{n-1}\mathbf{u}),$$

where in the last equality we have applied the Leibniz formula again. Consequently, using (7.13) and the second relation in (7.10), and taking into account the identities

$$(\psi_n'\phi + \psi_n\psi_{n-1})' = 2d_{2n-1}\psi_n, \quad (\psi_n'\phi + \psi_n\psi_{n-1})'' = 2d_{2n}d_{2n-1},$$

we deduce

$$\frac{(n-1)(n-2)}{2}D^{n-3}\left(\phi^{n-1}\mathbf{u}\right) \tag{7.14}$$

$$= \frac{1}{2d_{2n}d_{2n-1}}\Big(D^{n+1}(\phi^{n+1}\mathbf{u}) - (\psi_n'\phi + \psi_n\psi_{n-1})R_{n-1}\mathbf{u}$$

$$-\frac{2d_{2n-1}}{d_{2n-2}}\psi_n(R_n - \psi_{n-1}R_{n-1})\mathbf{u}\Big).$$

Now, consider the left-hand side of (7.12). By the Leibniz formula, we have

$$D^{n-1}(\phi^n\mathbf{u}) = \phi D^{n-1}(\phi^{n-1}\mathbf{u}) + (n-1)\phi' D^{n-2}(\phi^{n-1}\mathbf{u})$$

$$+ (n-1)(n-2)aD^{n-3}(\phi^{n-1}\mathbf{u}).$$

Substituting into (7.12), and using (7.13) and the second relation in (7.10), we deduce

$$\frac{1}{nd_{2n}}\left(D^{n+1}(\phi^{n+1}\mathbf{u}) - \psi_n R_n\mathbf{u}\right) \tag{7.15}$$

$$= \phi R_{n-1}\mathbf{u} + \frac{\phi'}{d_{2n-2}}(R_n - \psi_{n-1}R_{n-1})\mathbf{u} + a(n-1)(n-2)D^{n-3}(\phi^{n-1}\mathbf{u}).$$

Finally, substituting (7.14) in the right-hand side of (7.15), we obtain (7.11), where we have defined

$$\frac{d_{n-1}}{d_{2n-1}d_{2n}}R_{n+1}(x) := \left(x + \frac{d_{-2}q + 2bnd_{n-1}}{d_{2n-2}d_{2n}}\right)R_n(x)$$

$$+ \frac{nd_{2n-2}}{d_{2n-1}}\phi\left(-\frac{e_{n-1}}{d_{2n-2}}\right)R_{n-1}(x).$$

Since (by the induction hypothesis) R_n and R_{n-1} have degrees n and $n-1$, respectively, it follows that R_{n+1} is a polynomial of degree $n+1$, and the theorem is proved. □

7.2 Regularity Conditions

In the previous section we analyzed the functional differential equation (7.1) without requiring the regularity condition on the functional $\mathbf{u} \in \mathcal{P}'$. In this section we determine necessary and sufficient conditions, involving only the polynomials ϕ and ψ, which ensure the regularity of such a function. Notice that if both ϕ and ψ vanish identically, the equation (7.1) reduces to a trivial equation, so we will exclude this situation.

Lemma 7.2. *Suppose that $\mathbf{u} \in \mathcal{P}'$ is regular and satisfies the functional differential equation* (7.1), *where $\phi \in \mathcal{P}_2$ and $\psi \in \mathcal{P}_1$. Assume that at least one of the polynomials ϕ and ψ is nonzero. Then neither ϕ nor ψ is the zero polynomial and*

$$\deg \psi = 1. \tag{7.16}$$

Proof. Since $\mathbf{u}$ is regular, there is a monic OPS, $\{P_n\}_{n\geq 0}$, with respect to $\mathbf{u}$. Since $\mathbf{u}$ fulfills (7.1), if $\psi \equiv 0$ then $D(\phi\mathbf{u}) = \mathbf{0}$. If $\phi \not\equiv 0$, setting $r := \deg \phi$ and denoting by $k(\neq 0)$ the leading coefficient of ϕ, we have

$$\langle \mathbf{u}, P_r^2 \rangle = k^{-1}\langle \phi\mathbf{u}, P_r \rangle = -k^{-1}\Big\langle D(\phi\mathbf{u}), \int P_r \Big\rangle = -k^{-1}\Big\langle \mathbf{0}, \int P_r \Big\rangle = 0,$$

contrary to the regularity of $\mathbf{u}$. We conclude that $\psi \equiv 0$ implies $\phi \equiv 0$. Suppose now that $\phi \equiv 0$. Then, $\psi\mathbf{u} = 0$. If $\psi \not\equiv 0$, setting $t := \deg \psi$ and being $m(\neq 0)$ the leading coefficient of ψ, we have

$$\langle \mathbf{u}, P_t^2 \rangle = m^{-1}\langle \psi\mathbf{u}, P_t \rangle = 0,$$

contrary to the regularity of $\mathbf{u}$. We conclude that $\phi \equiv 0$ implies $\psi \equiv 0$. Finally, suppose that $\psi \equiv \text{constant} = q \neq 0$. Then

$$\langle \mathbf{u}, 1 \rangle = q^{-1}\langle \psi\mathbf{u}, 1 \rangle = q^{-1}\langle D(\phi\mathbf{u}), 1 \rangle = 0,$$

contrary to the regularity of $\mathbf{u}$, and so $\deg \psi = 1$. □

Given a monic polynomial R_n of degree n (which does not need to belong to an OPS), we denote by $R_n^{[k]}$ the monic polynomial of degree n defined by

$$R_n^{[k]}(x) := \frac{\mathrm{d}^k}{\mathrm{d}x^k} \frac{R_{n+k}(x)}{(n+1)_k}, \quad k \in \mathbb{N}_0, \tag{7.17}$$

where, for a given $\alpha \in \mathbb{C}$, $(\alpha)_n$ is the *Pochhammer symbol*, defined as

$$(\alpha)_0 := 1, \qquad (\alpha)_n := \alpha(\alpha+1)\cdots(\alpha+n-1), \quad n \in \mathbb{N}. \tag{7.18}$$

Clearly, if $\{R_n\}_{n\geq 0}$ is a simple set in $\mathcal{P}$, then so is $\{R_n^{[k]}\}_{n\geq 0}$. Under such conditions, there is a beautiful relation between their associated dual bases.

Lemma 7.3. *Let $\{R_n\}_{n\geq 0}$ be a simple set in $\mathcal{P}$ and let $\{\mathbf{a}_n\}_{n\geq 0}$ and $\left\{\mathbf{a}_n^{[k]}\right\}_{n\geq 0}$, $k \in \mathbb{N}_0$, be the dual bases in $\mathcal{P}'$ associated with $\{R_n\}_{n\geq 0}$ and $\{R_n^{[k]}\}_{n\geq 0}$, respectively. Then,*

$$D^k\left(\mathbf{a}_n^{[k]}\right) = (-1)^k (n+1)_k\, \mathbf{a}_{n+k}.$$

Proof. Fix $j, k, n \in \mathbb{N}_0$. It is easy to check that

$$\begin{aligned}\left\langle D^k\left(\mathbf{a}_n^{[k]}\right), R_j \right\rangle &= (-1)^k \left\langle \mathbf{a}_n^{[k]}, \frac{\mathrm{d}^k}{\mathrm{d}x^k} R_j \right\rangle \\ &= (-1)^k (j-k+1)_k \left\langle \mathbf{a}_n^{[k]}, R_{j-k}^{[k]} \right\rangle \\ &= (-1)^k (n+1)_k\, \delta_{n,j-k} = \left\langle (-1)^k (n+1)_k\, \mathbf{a}_{n+k}, R_j \right\rangle,\end{aligned}$$

and the result follows. □

Lemma 7.4. *Suppose that $\mathbf{u} \in \mathcal{P}'$ satisfies the functional differential equation (7.1), with ϕ and ψ given by (7.2), being at least one of these polynomials nonzero. Suppose further that $\mathbf{u}$ is regular. Then*

$$d_n := na + p \neq 0, \quad n \in \mathbb{N}_0. \tag{7.19}$$

Moreover, let $\{P_n\}_{n\geq 0}$ be the monic OPS with respect to $\mathbf{u}$ and let $P_n^{[k]}$ be defined by (7.17). Then, for each $k \in \mathbb{N}_0$, $\mathbf{u}^{[k]} := \phi^k \mathbf{u}$ is regular and $\{P_n^{[k]}\}_{n\geq 0}$ is the corresponding monic OPS.

Proof. By Lemma 7.2, both ϕ and ψ are nonzero, and $\deg \psi = 1$, and so $p \neq 0$. Consider first the case $k = 1$. Write

$$Q_n = P_n^{[1]} := \frac{P'_{n+1}}{n+1}.$$

We will show that

$$\langle \phi\mathbf{u}, Q_n Q_m\rangle = -\frac{d_n}{n+1}\langle \mathbf{u}, P_{n+1}^2\rangle \delta_{nm}, \quad n,m \in \mathbb{N}_0. \tag{7.20}$$

Indeed, since $D(\phi\mathbf{u}) = \psi\mathbf{u}$, we may write

$$\begin{aligned}(n+1)\langle \phi\mathbf{u}, Q_m Q_n\rangle &= \langle \phi\mathbf{u}, Q_m P_{n+1}'\rangle = \langle \phi\mathbf{u}, (Q_m P_{n+1})' - Q_m' P_{n+1}\rangle \\ &= -\langle D(\phi\mathbf{u}), Q_m P_{n+1}\rangle - \langle \phi\mathbf{u}, Q_m' P_{n+1}\rangle \\ &= -\langle \mathbf{u}, (\psi Q_m + \phi Q_m') P_{n+1}\rangle.\end{aligned}$$

Assume, without loss of generality, $m \leq n$. Since $\deg(\psi Q_m + \phi Q_m') < n+1$ if $m < n$ and

$$\psi Q_m + \phi Q_m' = (na+p)x^{n+1} + \pi_n(x)$$

if $m = n$, where $\pi_n \in \mathcal{P}_n$, the equation (7.20) follows. Let $s := \deg\phi \in \{0,1,2\}$. For each $n \in \mathbb{N}_0$, write

$$P_{n+s}(x) = \sum_{m=0}^{n+s} a_{nm} Q_m(x), \quad a_{nm} \in \mathbb{C}.$$

Multiplying both sides of this equality by ϕQ_n and then applying $\mathbf{u}$, and taking into account (7.20), we deduce

$$\langle \mathbf{u}, \phi Q_n P_{n+s}\rangle = -a_{nn} d_n \frac{\langle \mathbf{u}, P_{n+1}^2\rangle}{n+1}.$$

The left-hand side of this equality never vanishes, because $\{P_n\}_{n\geq 0}$ is an OPS with respect to $\mathbf{u}$ and $\deg\phi = s$ (and ϕ is not the zero polynomial), and so the right-hand side of the equality cannot vanish and $d_n \neq 0$ (and also $a_{nn} \neq 0$), which proves (7.19).

It remains to prove that $\{P_n^{[k]}\}_{n\geq 0}$ is a monic OPS with respect to $\mathbf{u}^{[k]} := \phi^k\mathbf{u}$, for each $k \in \mathbb{N}$. From (7.20) for $k=1$, we have

$$\left\langle \mathbf{u}^{[1]}, P_n^{[1]} P_m^{[1]}\right\rangle = -\frac{na+p}{n+1}\left\langle \mathbf{u}, P_{n+1}^2\right\rangle \delta_{nm}, \quad m \in \mathbb{N}_0, \tag{7.21}$$

and hence (7.19) ensures that $\{P_n^{[1]}\}_{n\geq 0}$ is a monic OPS with respect to $\mathbf{u}^{[1]} := \phi\mathbf{u}$. Now, by Lemma 7.1, $\mathbf{u}^{[1]}$ fulfills the functional differential equation $D\left(\phi\mathbf{u}^{[1]}\right) = \psi_1\mathbf{u}^{[1]}$. Since

$$P_n^{[2]} = \frac{\left(P_{n+1}^{[1]}\right)'}{n+1}, \quad \psi_1(x) = (2a+p)x + b + q,$$

from (7.21) with $\mathbf{u}$, ψ, and $\{P_n\}_{n\geq 0}$ replaced by $\mathbf{u}^{[1]}$, ψ_1, and $\{P_n^{[1]}\}_{n\geq 0}$, respectively, we get

$$\left\langle \mathbf{u}^{[2]}, P_n^{[2]} P_m^{[2]} \right\rangle = -\frac{(n+2)a+p}{n+1} \left\langle \mathbf{u}^{[1]}, \left(P_{n+1}^{[1]}\right)^2 \right\rangle \delta_{nm}$$

$$= \frac{d_{n+1}d_{n+2}}{(n+1)(n+2)} \langle \mathbf{u}, P_{n+2}^2 \rangle \delta_{nm}.$$

Therefore, $\{P_n^{[2]}\}_{n\geq 0}$ is a monic OPS with respect to $\mathbf{u}^{[2]}$. By induction on k we can prove that

$$\left\langle \mathbf{u}^{[k]}, P_n^{[k]} P_m^{[k]} \right\rangle = \frac{(-1)^k}{(n+1)_k} \prod_{j=0}^{k-1} d_{n+k+j-1} \langle \mathbf{u}, P_{n+k}^2 \rangle \delta_{nm}, \quad k, m \in \mathbb{N}_0,$$

and the result follows. □

We may now establish necessary and sufficient conditions ensuring the regularity of a given functional $\mathbf{u} \in \mathcal{P}'$ satisfying (7.2).

Theorem 7.2. *Suppose that $\mathbf{u} \in \mathcal{P}'$ satisfies the functional differential equation (7.1), with ϕ and ψ given by (7.2), where at least one of these polynomials is nonzero. Set*

$$d_n := na + p, \quad e_n := nb + q.$$

Then, $\mathbf{u}$ is regular if and only if

$$d_n \neq 0, \quad \phi\left(-\frac{e_n}{d_{2n}}\right) \neq 0, \quad n \in \mathbb{N}_0. \tag{7.22}$$

Moreover, under these conditions, the monic OPS $\{P_n\}_{n\geq 0}$ with respect to $\mathbf{u}$ is given by the recurrence relation

$$P_{n+1}(x) = (x - \beta_n)P_n(x) - \gamma_n P_{n-1}(x), \tag{7.23}$$

with initial conditions $P_{-1}(x) = 0$ and $P_0(x) = 1$, where

$$\beta_n = \frac{ne_{n-1}}{d_{2n-2}} - \frac{(n+1)e_n}{d_{2n}}, \quad n \in \mathbb{N}_0, \tag{7.24}$$

$$\gamma_{n+1} = -\frac{(n+1)d_{n-1}}{d_{2n-1}d_{2n+1}} \phi\left(-\frac{e_n}{d_{2n}}\right), \quad n \in \mathbb{N}. \tag{7.25}$$

In addition, P_n satisfies the functional Rodrigues formula

$$P_n \mathbf{u} = k_n D^n (\phi^n \mathbf{u}), \quad k_n := \prod_{i=0}^{n-1} \frac{1}{d_{n+i-1}}. \tag{7.26}$$

Proof. Suppose that $\mathbf{u}$ is regular. Let $\{P_n\}_{n\geq 0}$ be the monic OPS with respect to $\mathbf{u}$. By Lemma 7.4, $d_n \neq 0$ for each $n \in \mathbb{N}_0$. Moreover, $\mathbf{u}^{[k]} := \phi^k \mathbf{u}$ is regular and $\{P_n^{[k]}\}_{n\geq 0}$ is its monic OPS for each $k \in \mathbb{N}_0$. Note that $\{P_n^{[k]}\}_{n\geq 0}$ satisfies

$$P_{n+1}^{[k]}(x) = (x - \beta_n^{[k]})P_n^{[k]}(x) - \gamma_n^{[k]} P_{n-1}^{[k]}(x), \tag{7.27}$$

with initial conditions $P_{-1}^{[k]}(x) = 0$ and $P_0^{[k]}(x) = 1$, where $\beta_n^{[k]} \in \mathbb{C}$ and $\gamma_n^{[k]} \in \mathbb{C} \setminus \{0\}$. Let us compute $\gamma_1^{[n]}$ for each fixed $n \in \mathbb{N}_0$. We first show that the coefficient $\gamma_1 \equiv \gamma_1^{[0]}$, appearing in the recurrence relation for $\{P_n\}_{n\geq 0}$, is given by

$$\gamma_1 = -\frac{1}{p+a}\phi\left(-\frac{q}{p}\right). \tag{7.28}$$

To prove this relation, take $n = 0$ and $n = 1$ in the recurrence relation (9.80) for the sequence of moments associated to $\mathbf{u}$, which gives

$$u_1 = -\frac{q}{p}u_0, \quad u_2 = -\frac{1}{p+a}\left(-(b+q)\frac{q}{p} + c\right)u_0. \tag{7.29}$$

On the other hand, by Corollary 2.4, we get

$$\gamma_1 = \frac{H_{-1}H_1}{H_0^2} = \frac{u_2u_0 - u_1^2}{u_0^2}. \tag{7.30}$$

Substituting u_1 and u_2 given by (7.29) into (7.30) yields (7.28). Now, since equation (7.6) is of the same type as equation (7.1), with the same polynomial ϕ and being ψ replaced by ψ_n, we see that the expression of $\gamma_1^{[n]}$ may be obtained replacing the coefficients p and q of ψ in (7.28) by the corresponding coefficients of ψ_n. Hence,

$$\gamma_1^{[n]} = -\frac{1}{d_{2n}+a}\phi\left(-\frac{e_n}{d_{2n}}\right) = -\frac{1}{d_{2n+1}}\phi\left(-\frac{e_n}{d_{2n}}\right). \tag{7.31}$$

Since $\mathbf{u}^{[n]}$ is regular, then $\gamma_1^{[n]} \neq 0$. Thus, the second condition in (7.22) holds.

Conversely, suppose that conditions (7.22) hold. According with Theorem 7.1, there is a simple set of polynomials $\{R_n\}_{n\geq 0}$ satisfiying Theorem 7.1. The hypothesis (7.22) ensures that $\widetilde{\alpha}_n \neq 0$ and $\widetilde{\gamma}_n \neq 0$ for each n. Thus, by Favard's theorem, $\{R_n\}_{n\geq 0}$ is an OPS. We claim that $\{R_n\}_{n\geq 0}$ is an OPS with respect to $\mathbf{u}$. By Corollary 6.1, we only need to show that

$$u_0 \neq 0, \quad \langle \mathbf{u}, R_n \rangle = 0. \tag{7.32}$$

In fact, if $u_0 = 0$, since (by Lemma 7.1) the functional differential equation (7.1) is equivalent to the recurrence relation (9.80) fulfilled by the moments u_n, and since for $n = 0$ (9.80) yields $pu_1 + qu_0 = 0$, we would get $pu_1 = 0$ and so $u_1 = 0$ (because $p = d_0 \neq 0$); therefore, $u_0 = u_1 = 0$, and it follows recurrently from (9.80) that $u_n = 0$ for each $n \in \mathbb{N}_0$, hence $\mathbf{u} = \mathbf{0}$, in contradiction with the hypothesis. Hence $u_0 \neq 0$. On the other hand, by (7.8), we have

$$\langle \mathbf{u}, R_n \rangle = \langle R_n \mathbf{u}, 1 \rangle = (-1)^n \langle \phi^n \mathbf{u}, 0 \rangle = 0.$$

Thus (7.32) is proved. Therefore, $\{R_n\}_{n \geq 0}$ is a monic OPS with respect to $\mathbf{u}$, hence $\mathbf{u}$ is regular. It remains to prove (7.24)–(7.26). Since $\{P_n\}_{n \geq 0}$ and $\{R_n\}_{n \geq 0}$ are both OPS with respect to $\mathbf{u}$, then there exists a sequence $\{k_n\}_{n \geq 0}$, with $k_n \in \mathbb{C} \setminus \{0\}$, such that

$$P_n(x) = k_n R_n(x). \tag{7.33}$$

Multiplying both sides of (7.9) by k_n we obtain

$$xP_n(x) = \frac{k_n}{\widetilde{\alpha}_n k_{n+1}} P_{n+1}(x) + \frac{\widetilde{\beta}_n}{\widetilde{\alpha}_n} P_n(x) + \frac{\widetilde{\gamma}_n k_n}{\widetilde{\alpha}_n k_{n-1}} P_{n-1}(x),$$

and $P_1 = x - \widetilde{\beta}_0$ where $\widetilde{\beta}_0 = -q/p$. Since each P_n is a monic polynomial, we must have

$$\frac{1}{\widetilde{\alpha}_n} \frac{k_n}{k_{n+1}} = 1.$$

Therefore, since $k_0 = 1$, it follows that

$$k_n = \prod_{i=0}^{n-1} \frac{1}{\widetilde{\alpha}_i} = \prod_{i=0}^{n-1} \frac{1}{d_{n+i-1}}.$$

Thus (7.26) follows from (7.33) and (7.8). Finally, by Theorem 7.1, the coefficients of the recurrence relation for $\{P_n\}_{n \geq 0}$ are given by

$$\beta_n = \frac{\widetilde{\beta}_n}{\widetilde{\alpha}_n} = -\frac{d_{-2} q + 2bn d_{n-1}}{d_{2n} d_{2n-2}},$$

$$\gamma_n = \frac{\widetilde{\gamma}_n k_n}{\widetilde{\alpha}_n k_{n-1}}, = -\frac{n d_{n-2}}{d_{2n-3} d_{2n-1}} \phi\left(-\frac{e_{n-1}}{d_{2n-2}}\right),$$

which completes the proof. □

Remark 7.1. *The regularity conditions (7.22) may be expressed as*

$$d_n \left(a e_n^2 - b e_n d_{2n} + c d_{2n}^2\right) \neq 0, \quad n \in \mathbb{N}_0. \tag{7.34}$$

Exercises

1. Let $\mathbf{u} \equiv \mathbf{u}(r_1, r_2) \in \mathcal{P}'$, $r_1, r_2 \in \mathbb{C}$, be a solution of the functional differential equation
$$D\left((x-r_1)(x-r_2)\mathbf{u}\right) = \left(x - \frac{r_1+r_2}{2}\right)\mathbf{u}.$$
   (a) Prove that $\mathbf{u}$ is regular if and only if $r_1 \neq r_2$.

   (b) Assuming $r_1 \neq r_2$, show that the monic OPS, $\{P_n\}_{n\geq 0}$, with respect to $\mathbf{u}$ is given by
$$P_n(x) := \left(\frac{r_1-r_2}{4}\right)^n U_n\left(\frac{2x-r_1-r_2}{r_1-r_2}\right),$$
   $\{U_n\}_{n\geq 0}$ being the Chebyshev polynomials of the second kind.

Chapter 8

Classical Orthogonal Polynomials: General Properties

The classical functionals are the regular solutions of the functional differential equation $D(\phi\mathbf{u}) = \psi\mathbf{u}$, where $\phi \in \mathcal{P}_2$ and $\psi \in \mathcal{P}_1$ are nonzero polynomials. The corresponding orthogonal polynomial sequences are called classical orthogonal polynomials, namely Hermite, Laguerre, Jacobi, and Bessel polynomials. In this text we present the most significant results concerning this important class of orthogonal polynomials.

8.1 Definition and Characterizations

We begin this chapter with the definitions of classical functionals and classical OPS.

Definition 8.1. *The functional* $\mathbf{u} \in \mathcal{P}'$ *is called a classical functional if the following two conditions hold:*

(i) $\mathbf{u}$ *is regular;*

(ii) $\mathbf{u}$ *satisfies the functional differential equation*

$$D(\phi\mathbf{u}) = \psi\mathbf{u}, \tag{8.1}$$

where ϕ *and* ψ *are polynomials fulfilling*

$$\deg\phi \leq 2, \quad \deg\psi = 1. \tag{8.2}$$

An OPS $\{P_n\}_{n\geq 0}$ *with respect to a classical functional is called a classical OPS.*

DOI: 10.1201/9781003432067-8

Remark 8.1. *According to Lemma 7.2, the conditions (8.2) may be replaced by the weaker conditions*

$$\phi \in \mathcal{P}_2, \quad \psi \in \mathcal{P}_1, \quad \{\phi, \psi\} \neq \mathcal{P}_{-1} := \{0\}. \tag{8.3}$$

Theorem 7.2 gives necessary and sufficient conditions for the existence of solutions of the functional differential equation, characterizing also such functionals (in particular, solving the question of the existence of classical functionals). Thus, we may state: *a functional* $\mathbf{u} \in \mathcal{P}'$ *is classical if and only if there exist* $\phi \in \mathcal{P}_2$ *and* $\psi \in \mathcal{P}_1$ *such that the following conditions hold:*

$$\begin{aligned} &\text{(i)} \quad D(\phi\mathbf{u}) = \psi\mathbf{u}; \\ &\text{(ii)} \quad na + p \neq 0, \quad \phi\left(-\frac{nb+q}{2na+p}\right) \neq 0, \quad n \in \mathbb{N}_0, \end{aligned} \tag{8.4}$$

where we have set $\phi(x) = ax^2 + bx + c$ *and* $\psi(x) = px + q$.

In the next theorem we state several characterizations of the classical OPS. For convenience, we introduce the concept of admissible pair.

Definition 8.2. (ϕ, ψ) *is called an admissible pair if*

$$\phi \in \mathcal{P}_2, \quad \psi \in \mathcal{P}_1, \quad d_n := \psi' + \frac{n}{2}\phi'' \neq 0, \quad n \in \mathbb{N}_0.$$

Introducing this concept makes sense, since according to conditions (ii) in (8.4), only admissible pairs may appear in the framework of the theory of classical OP.

Theorem 8.1 (Characterization Theorem). *Let* $\mathbf{u} \in \mathcal{P}'$ *be regular and let* $\{P_n\}_{n\geq 0}$ *be the corresponding monic OPS. Then the following properties are equivalent:*

C1. $\mathbf{u}$ *is classical, i.e., there are nonzero polynomials* $\phi \in \mathcal{P}_2$ *and* $\psi \in \mathcal{P}_1$ *such that* $\mathbf{u}$ *satisfies the functional differential equation*

$$D(\phi\mathbf{u}) = \psi\mathbf{u};$$

C1′. *there is an admissible pair* (ϕ, ψ) *such that* $\mathbf{u}$ *satisfies the distributional differential equation*

$$D(\phi\mathbf{u}) = \psi\mathbf{u};$$

C2. (Al-Salam-Chihara) *there exists a polynomial* $\phi \in \mathcal{P}_2$ *and complex parameters* a_n, b_n, *and* c_n, *with* $c_n \neq 0$ *for each* $n \in \mathbb{N}$, *such that*

$$\phi(x)P_n'(x) = a_n P_{n+1}(x) + b_n P_n(x) + c_n P_{n-1}(x);$$

C3. (Hahn) $\left\{ P_n^{[k]} := \frac{\mathrm{d}^k}{\mathrm{d}x^k} \frac{P_{n+k}}{(n+1)_k} \right\}_{n\geq 0}$ *is a monic OPS for some* $k \in \mathbb{N}$*;*

C3′. $\left\{ P_n^{[k]} \right\}_{n\geq 0}$ *is a monic OPS for each* $k \in \mathbb{N}$*;*

C4. *there exist* $k \in \mathbb{N}$ *and complex parameters* $r_n^{[k]}$ *and* $s_n^{[k]}$ *such that*

$$P_n^{[k-1]}(x) = P_n^{[k]}(x) + r_n^{[k]} P_{n-1}^{[k]}(x) + s_n^{[k]} P_{n-2}^{[k]}(x), \quad n \in \mathbb{N} \setminus \{1\};$$

C4′. *for each* $k \in \mathbb{N}$, *there exist parameters* $r_n^{[k]}$ *and* $s_n^{[k]}$ *such that C4 holds;*

C5. (Bochner) *there exist polynomials* ϕ *and* ψ *and, for each* $n \in \mathbb{N}_0$, *a complex parameter* λ_n, *with* $\lambda_n \neq 0$ *for each* $n \in \mathbb{N}$, *such that* $y = P_n(x)$ *is a solution of the second-order ordinary differential equation*

$$\phi(x)y'' + \psi(x)y' + \lambda_n y = 0;$$

C6. (Maroni) *there is an admissible pair* (ϕ, ψ) *so that the formal Stieltjes series associated with* $\mathbf{u}$,

$$S_{\mathbf{u}}(z) := -\sum_{n=0}^{\infty} u_n / z^{n+1},$$

satisfies (formally)

$$\phi(z)S_{\mathbf{u}}'(z) = (\psi(z) - \phi'(z))S_{\mathbf{u}}(z) + (\psi' - \tfrac{1}{2}\phi'')u_0;$$

C7. (McCarthy) *there exists an admissible pair* (ϕ, ψ) *and, for each* $n \in \mathbb{N}$, *complex parameters* h_n *and* t_n *such that*

$$\phi(P_n P_{n-1})'(x) = h_n P_n^2(x) - (\psi - \phi')P_n P_{n-1}(x) + t_n P_{n-1}^2(x);$$

C8. (Functional Rodrigues Formula) *there exists a polynomial* $\phi \in \mathcal{P}_2$ *and nonzero complex parameters* k_n *such that*

$$P_n(x)\mathbf{u} = k_n D^n \left(\phi^n(x)\mathbf{u} \right).$$

In addition, let the three-term recurrence relation fulfilled by the monic OPS $\{P_n\}_{n\geq 0}$ *be*

$$P_{n+1}(x) = (x - \beta_n)P_n(x) - \gamma_n P_{n-1}(x),$$

with initial conditions $P_{-1}(x) = 0$ *and* $P_0(x) = 1$. *Write* $\phi(x) = ax^2 + bx + c$, $\psi(x) = px + q$, $d_n := na + p$, *and* $e_n := nb + q$. *Then*

$$\beta_n = -\frac{d_{-2}q + 2bnd_{n-1}}{d_{2n}d_{2n-2}},$$

$$\gamma_n = -\frac{nd_{n-2}}{d_{2n-3}d_{2n-1}}\phi\left(-\frac{e_{n-1}}{d_{2n-2}}\right),$$

and the parameters appearing in the above characterizations may be computed explicitly:

$$a_n = na, \qquad b_n = -\frac{1}{2}\psi(\beta_n), \qquad c_n = -d_{n-1}\gamma_n,$$

$$r_n^{[1]} = \frac{1}{2}\frac{\psi(\beta_n)}{d_{n-1}}, \qquad s_n^{[1]} = -\frac{(n-1)a}{d_{n-2}}\gamma_n, \qquad \lambda_n = -nd_{n-1},$$

$$h_n = d_{2n-3}, \qquad t_n = -d_{2n-1},\gamma_n, \qquad k_n = \prod_{i=0}^{n-1} d_{n+i-1}^{-1}.$$

Proof. By Lemma 7.4 and Theorem 7.2, it is easy to see that C1 ⇔ C1′, C1 ⇒ C3′, and C1′ ⇔ C8. Clearly, C3′ ⇒ C3 and C4′ ⇒ C4. We show that C3′ ⇒ C4′ using the same arguments of the proof of C3 ⇒ C4 given below. The proof of C1′ ⇔ C6 is left to the reader. Consequently, we only need to show that:

$$\text{C1}' \Rightarrow \text{C2} \Rightarrow \text{C3} \Rightarrow \text{C4} \Rightarrow \text{C1}, \quad \text{C1} \Leftrightarrow \text{C5}, \quad \text{C2} \Leftrightarrow \text{C7}.$$

(C1′ ⇒ C2). Assume that C1′ holds. Since $\deg(\phi P_n') \leq n+1$, we have

$$\phi P_n' = \sum_{j=0}^{n+1} a_{n,j}P_j, \quad a_{n,j} := \frac{\langle \mathbf{u}, \phi P_n' P_j\rangle}{\langle \mathbf{u}, P_j^2\rangle}. \tag{8.5}$$

For each integer number j, with $0 \leq j \leq n+1$, we deduce

$$\begin{aligned}\langle \mathbf{u}, \phi P_n' P_j\rangle &= \langle \phi\mathbf{u}, (P_nP_j)' - P_nP_j'\rangle \\ &= -\langle D(\phi\mathbf{u}), P_nP_j\rangle - \langle \phi\mathbf{u}, P_nP_j'\rangle \\ &= -\langle \mathbf{u}, \psi P_jP_n\rangle - \langle \mathbf{u}, \phi P_j' P_n\rangle.\end{aligned} \tag{8.6}$$

If $0 \leq j \leq n-2$, we obtain $\langle \mathbf{u}, \phi P_n' P_j \rangle = 0$, and so $a_{n,j} = 0$. Thus, (8.5) reduces to

$$\phi P_n' = a_n P_{n+1} + b_n P_n + c_n P_{n-1},$$

where, writing $\phi(x) = ax^2 + bx + c$ and $\psi(x) = px + q$, $a_n = na$ (by comparison of coefficients), $b_n = a_{n,n}$, and $c_n := a_{n,n-1}$. Setting $j = n-1$ in (8.6), we deduce

$$\langle \mathbf{u}, \phi P_n' P_{n-1} \rangle = -\langle \mathbf{u}, (\psi P_{n-1} + \phi P_{n-1}') P_n \rangle = -d_{n-1} \langle \mathbf{u}, P_n^2 \rangle.$$

Hence

$$c_n := a_{n,n-1} = \frac{\langle \mathbf{u}, \phi P_n' P_{n-1} \rangle}{\langle \mathbf{u}, P_{n-1}^2 \rangle}$$

$$= \frac{\langle \mathbf{u}, \phi P_n' P_{n-1} \rangle}{\langle \mathbf{u}, P_n^2 \rangle} \frac{\langle \mathbf{u}, P_n^2 \rangle}{\langle \mathbf{u}, P_{n-1}^2 \rangle} = -d_{n-1} \gamma_n, \quad n \in \mathbb{N}.$$

Since, by hypothesis, (ϕ, ψ) is an admissible pair, then we may conclude that $c_n \neq 0$, and so C1′ ⇒ C2. Taking $j = n$ in (8.6) yields

$$\langle \mathbf{u}, \phi P_n' P_n \rangle = -\frac{1}{2} \langle \mathbf{u}, \psi P_n^2 \rangle = -\frac{1}{2} p \langle \mathbf{u}, x P_n^2 \rangle - \frac{1}{2} q \langle \mathbf{u}, P_n^2 \rangle,$$

hence we deduce the expression for b_n given in the statement of the theorem:

$$b_n := a_{n,n} = \frac{\langle \mathbf{u}, \phi P_n' P_n \rangle}{\langle \mathbf{u}, P_n^2 \rangle}$$

$$= -\frac{1}{2} p \frac{\langle \mathbf{u}, x P_n^2 \rangle}{\langle \mathbf{u}, P_n^2 \rangle} - \frac{1}{2} q = -\frac{1}{2} \psi(\beta_n).$$

(C2 ⇒ C3). Suppose that C2 holds. We claim that

$$\{P_n^{[1]} := P_{n+1}'/(n+1)\}_{n \geq 0}$$

is a monic OPS with respect to $\mathbf{v} := \phi \mathbf{u}$. Indeed, for each $n \in \mathbb{N}_0$ and $0 \leq m \leq n$, we have

$$(n+1) \langle \mathbf{v}, x^m P_n^{[1]} \rangle = \langle \phi \mathbf{u}, x^m P_{n+1}' \rangle = \langle \mathbf{u}, \left(\phi P_{n+1}' \right) x^m \rangle$$

$$= \langle \mathbf{u}, (a_{n+1} P_{n+2} + b_{n+1} P_{n+1} + c_{n+1} P_n) x^m \rangle$$

$$= c_{n+1} \langle \mathbf{u}, P_n^2 \rangle \delta_{m,n}.$$

Since (by hypothesis) $c_{n+1} \neq 0$, we conclude that $\{P_n^{[1]}\}_{n \geq 0}$ is a monic OPS (with respect to $\mathbf{v} := \phi \mathbf{u}$) as claimed.

(C3 ⇒ C4). By hypothesis,

$$\left\{ P_n^{[k]} := \frac{\mathrm{d}^k}{\mathrm{d}x^k} \left(\frac{P_{n+k}}{(n+1)_k} \right) \right\}_{n \geq 0}$$

is a monic OPS for some (fixed) $k \in \mathbb{N}$. Then there exists $\beta_n^{[k]} \in \mathbb{C}$ and $\gamma_n^{[k]} \in \mathbb{C} \setminus \{0\}$ such that

$$xP_n^{[k]} = P_{n+1}^{[k]} + \beta_n^{[k]} P_n^{[k]} + \gamma_n^{[k]} P_{n-1}^{[k]}. \tag{8.7}$$

Similarly, there exist $\beta_n \in \mathbb{C}$ and $\gamma_n \in \mathbb{C} \setminus \{0\}$ such that

$$xP_n = P_{n+1} + \beta_n P_n + \gamma_n P_{n-1}. \tag{8.8}$$

Changing n into $n+k$ in (8.8), taking the derivative of order k on both sides of the resulting equation and using the Leibnitz formula on the left-hand side, we get

$$xP_n^{[k]} + \frac{k}{n+1} P_{n+1}^{[k-1]} = \frac{n+k+1}{n+1} P_{n+1}^{[k]} + \beta_{n+k} P_n^{[k]} + \frac{n\gamma_{n+k}}{n+k} P_{n-1}^{[k]}.$$

Now, replacing $xP_n^{[k]}$ by the right-hand side of (8.7), and then changing n into $n-1$, we obtain $C4$ with

$$r_n^{[k]} = \frac{n\left(\beta_{n+k-1} - \beta_{n-1}^{[k]}\right)}{k},$$

$$s_n^{[k]} = \frac{n\left((n-1)\gamma_{n+k-1} - (n+k-1)\gamma_{n-1}^{[k]}\right)}{k(n+k-1)}.$$

(C4 $\Rightarrow$ C1). By hypothesis C4 holds. Let $\{\mathbf{a}_n\}_{n\geq 0}$ and $\{\mathbf{a}_n^{[k]}\}_{n\geq 0}$ be the dual bases for $\{P_n\}_{n\geq 0}$ and $\{P_n^{[k]}\}_{n\geq 0}$, respectively. By Theorem 1.1,

$$\mathbf{a}_n^{[k]} = \sum_{j\geq 0} \langle \mathbf{a}_n^{[k]}, P_j^{[k-1]} \rangle \mathbf{a}_j^{[k-1]}.$$

From C4, we compute

$$\begin{aligned}\langle \mathbf{a}_n^{[k]}, P_j^{[k-1]} \rangle &= \langle \mathbf{a}_n^{[k]}, P_j^{[k]} \rangle + r_j^{[k]} \langle \mathbf{a}_n^{[k]}, P_{j-1}^{[k]} \rangle + s_j^{[k]} \langle \mathbf{a}_n^{[k]}, P_{j-2}^{[k]} \rangle \\ &= \begin{cases} 1, & \text{if } j = n, \\ r_{n+1}^{[k]}, & \text{if } j = n+1, \\ s_{n+2}^{[k]}, & \text{if } j = n+2, \\ 0, & \text{otherwise.} \end{cases}\end{aligned}$$

Hence

$$\mathbf{a}_n^{[k]} = \mathbf{a}_n^{[k-1]} + r_{n+1}^{[k]} \mathbf{a}_{n+1}^{[k-1]} + s_{n+2}^{[k]} \mathbf{a}_{n+2}^{[k-1]}.$$

Taking the (functional) derivative of order k in both sides of this equation, and using the relations

$$D^j \left(\mathbf{a}_n^{[j]}\right) = (-1)^j (n+1)_j \, \mathbf{a}_{n+j}$$

(see Lemma 7.3), we obtain

$$D\left(\frac{1}{n+k}\,\mathbf{a}_{n+k-1}+\frac{r_{n+1}^{[k]}}{n+1}\,\mathbf{a}_{n+k}+\frac{(n+k+1)s_{n+2}^{[k]}}{(n+1)(n+2)}\,\mathbf{a}_{n+k+1}\right)=-\mathbf{a}_{n+k}.$$

Therefore, since, by Theorem 6.1,

$$\mathbf{a}_j=\frac{P_j}{\langle \mathbf{u},P_j^2\rangle}\,\mathbf{u},\quad j\in\mathbb{N}_0$$

and, by Corollary 2.4,

$$\gamma_j=\langle \mathbf{u},P_j^2\rangle/\langle \mathbf{u},P_{j-1}^2\rangle,\quad j\in\mathbb{N},$$

γ_j being the γ−parameter appearing in (8.8), we deduce

$$D\left(\Phi_{n+k+1}\,\mathbf{u}\right)=-P_{n+k}\,\mathbf{u},\tag{8.9}$$

where Φ_{n+k+1} is a polynomial of degree at most $n+k+1$, given by

$$\Phi_{n+k+1}(x):=\frac{\gamma_{n+k}}{n+k}\,P_{n+k-1}(x)+\frac{r_{n+1}^{[k]}}{n+1}\,P_{n+k}(x)$$
$$+\frac{(n+k+1)s_{n+2}^{[k]}}{(n+1)(n+2)\gamma_{n+k+1}}\,P_{n+k+1}(x).$$

Since Φ_{n+k+1} is a (finite) linear combination of polynomials of the simple set $\{P_j\}_{j\geq 0}$ and $\gamma_{n+k}\neq 0$, then Φ_{n+k+1} does not vanish identically, so $\Phi_{n+k+1}\in\mathcal{P}_{n+k+1}\setminus\{0\}$. Setting $n=0$ and $n=1$ in (8.9) we obtain the two equations

$$D\left(\Phi_{k+1}\,\mathbf{u}\right)=-P_k\,\mathbf{u},\quad D\left(\Phi_{k+2}\,\mathbf{u}\right)=-P_{k+1}\,\mathbf{u}.\tag{8.10}$$

If $k=1$, it follows immediately from the first of these equations that C1 holds. Henceforth, assume that $k\in\mathbb{N}\setminus\{1\}$. Setting $n=0$ and $n=1$ in the definition of Φ_{n+k+1} and using the three-term recurrence relation (8.8), we easily deduce

$$\Phi_{k+1}(x)=E_0(x;k)P_{k+1}(x)+F_1(x;k)P_k(x),\tag{8.11}$$

$$\Phi_{k+2}(x)=G_1(x;k)P_{k+1}(x)+H_0(x;k)P_k(x),\tag{8.12}$$

where $E_0(\cdot;k),H_0(\cdot;k)\in\mathcal{P}_0$ and $F_1(\cdot;k),G_1(\cdot;k)\in\mathcal{P}_1$ are explicitly given by

$$E_0(x;k):=\frac{(k+1)s_2^{[k]}}{2\gamma_{k+1}}-\frac{1}{k},\quad F_1(x;k):=\frac{x-\beta_k}{k}+r_1^{[k]},\tag{8.13}$$

$$G_1(x;k):=\frac{(k+2)s_3^{[k]}(x-\beta_{k+1})}{6\gamma_{k+2}}+\frac{r_2^{[k]}}{2},\tag{8.14}$$

$$H_0(x;k):=\frac{\gamma_{k+1}}{k+1}-\frac{(k+2)s_3^{[k]}\gamma_{k+1}}{6\gamma_{k+2}}.\tag{8.15}$$

Set

$$\Delta_2(x) \equiv \Delta_2(x;k) := E_0(x;k)H_0(x;k) - F_1(x;k)G_1(x;k),$$

the determinant of the system given by equations (8.11) and (8.12). Using (8.10)–(8.15), and taking into account that $\mathbf{u}$ is regular, we prove that $\Delta_2 \in \mathcal{P}_2 \setminus \{0\}$. Solving (8.11) and (8.12) for P_k and P_{k+1} we obtain

$$\Delta_2(x)P_{k+1}(x) = H_0(x;k)\Phi_{k+1}(x) - F_1(x;k)\Phi_{k+2}(x), \tag{8.16}$$

$$\Delta_2(x)P_k(x) = E_0(x;k)\Phi_{k+2}(x) - G_1(x;k)\Phi_{k+1}(x). \tag{8.17}$$

Since P_k and P_{k+1} cannot share zeros, it follows from (8.16)–(8.17) that any common zero of Φ_{k+1} and Φ_{k+2} (if there is any) must be a zero of Δ_2. Let Φ be the greatest common divisor of Φ_{k+1} and Φ_{k+2}, i.e.,

$$\Phi(x) := \text{g.c.d.}\,\{\Phi_{k+1}(x), \Phi_{k+2}(x)\}.$$

Any zero of Φ is also a zero of both Φ_{k+1} and Φ_{k+2}, and so it is a zero of Δ_2. Therefore, $\Phi \in \mathcal{P}_2 \setminus \{0\}$. (Notice that indeed $\Phi \not\equiv 0$, since $\Phi_{k+1} \not\equiv 0$ and $\Phi_{k+2} \not\equiv 0$.) Moreover, there exist polynomials $\Phi_{1,k}$ and $\Phi_{2,k}$, with no common zeros, such that

$$\Phi_{k+1} = \Phi\,\Phi_{1,k}, \quad \Phi_{k+2} = \Phi\,\Phi_{2,k}, \tag{8.18}$$

$$\Phi_{1,k} \in \mathcal{P}_{k+1-\ell} \setminus \{0\}, \quad \Phi_{2,k} \in \mathcal{P}_{k+2-\ell} \setminus \{0\}, \quad \ell := \deg \Phi \leq 2. \tag{8.19}$$

From (8.10) and (8.18) we deduce

$$\Phi_{1,k}D(\Phi\mathbf{u}) = -(P_k + \Phi_{1,k}'\Phi)\mathbf{u}, \quad \Phi_{2,k}D(\Phi\mathbf{u}) = -(P_{k+1} + \Phi_{2,k}'\Phi)\mathbf{u}. \tag{8.20}$$

Combining these two equations yields

$$\big(\Phi_{1,k}(P_{k+1} + \Phi_{2,k}'\Phi) - \Phi_{2,k}(P_k + \Phi_{1,k}'\Phi)\big)\,\mathbf{u} = \mathbf{0},$$

and so, since $\mathbf{u}$ is regular,

$$\Phi_{1,k}(P_{k+1} + \Phi_{2,k}'\Phi) = \Phi_{2,k}(P_k + \Phi_{1,k}'\Phi).$$

Therefore, taking into account that $\Phi_{1,k}$ and $\Phi_{2,k}$ have no common zeros and (8.19) holds, we may ensure that there exists a polynomial $\Psi \in \mathcal{P}_1$ such that

$$P_k + \Phi_{1,k}'\Phi = -\Psi\Phi_{1,k}, \quad P_{k+1} + \Phi_{2,k}'\Phi = -\Psi\Phi_{2,k}. \tag{8.21}$$

Combining equations (8.20) and (8.21) we deduce

$$\Phi_{1,k}\,(D(\Phi\mathbf{u}) - \Psi\mathbf{u}) = \Phi_{2,k}\,(D(\Phi\mathbf{u}) - \Psi\mathbf{u}) = \mathbf{0}.$$

From these equations, and using once again the fact that $\Phi_{1,k}$ and $\Phi_{2,k}$ have no common zeros, we conclude, by Theorem 1.5, that $D(\Phi\mathbf{u}) = \Psi\mathbf{u}$. Thus

C4 ⇒ C1. The formulas for $r_n^{[1]}$ and $s_n^{[1]}$ given in the statement of the theorem may be derived as follows. We have already proved that C4 ⇒ C1 ⇒ C1′ ⇒ C2 ⇒ C3 ⇒ C4. As we have seen, the formulas for b_n and c_n given in the statement of the theorem hold. We now use these formulas to obtain the expressions for $r_n^{[1]}$ and $s_n^{[1]}$. Set $Q_n := P_n^{[1]} := P'_{n+1}/(n+1)$. Note that C4 yields

$$P_n = Q_n + r_n^{[1]} Q_{n-1} + s_n^{[1]} Q_{n-2}, \quad n \in \mathbb{N} \setminus \{1\}.$$

Hence, since $\{Q_n\}_{n \geq 0}$ is a monic OPS with respect to $\mathbf{v} := \phi \mathbf{u}$, we deduce, for each $n \in \mathbb{N} \setminus \{1\}$,

$$r_n^{[1]} = \frac{\langle \mathbf{u}, \phi P_n P'_n \rangle}{\langle \mathbf{u}, \phi P'_n P_{n-1} \rangle} = \frac{\langle \mathbf{u}, P_{n-1}^2 \rangle}{\langle \mathbf{u}, \phi P'_n P_{n-1} \rangle} \frac{\langle \mathbf{u}, \phi P'_n P_n \rangle}{\langle \mathbf{u}, P_n^2 \rangle} \frac{\langle \mathbf{u}, P_n^2 \rangle}{\langle \mathbf{u}, P_{n-1}^2 \rangle}$$

$$= \frac{1}{c_n} b_n \gamma_n = \frac{1}{2} \frac{\psi(\beta_n)}{d_{n-1}},$$

where the third equality holds taking into account C2. Similarly, for each $n \in \mathbb{N} \setminus \{1\}$,

$$s_n^{[1]} = \frac{a \langle \mathbf{u}, P_n^2 \rangle}{\dfrac{1}{n-1} \langle \mathbf{u}, \phi P'_{n-1} P_{n-2} \rangle} = \frac{(n-1) a \langle \mathbf{u}, P_n^2 \rangle}{c_{n-1} \langle \mathbf{u}, P_{n-2}^2 \rangle}$$

$$= \frac{(n-1)a}{c_{n-1}} \gamma_{n-1} \gamma_n = -\frac{(n-1)a}{d_{n-2}} \gamma_n.$$

(C1 ⇒ C5). By hypothesis, $D(\phi \mathbf{u}) = \psi \mathbf{u}$, where $\phi \in \mathcal{P}_2$, $\psi \in \mathcal{P}_1$, and $\deg \psi = 1$ (cf. Lemma 7.2). Fix $n \in \mathbb{N}$ and write

$$\phi P''_n + \psi P'_n = \sum_{j=0}^{n} \lambda_{n,j} P_j. \tag{8.22}$$

Then, for each j such that $0 \leq j \leq n$,

$$\langle \mathbf{u}, P_j^2 \rangle \lambda_{n,j} = \langle \mathbf{u}, (\phi P''_n + \psi P'_n) P_j \rangle = \langle \phi \mathbf{u}, P''_n P_j \rangle + \langle \psi \mathbf{u}, P'_n P_j \rangle$$

$$= \langle \phi \mathbf{u}, (P'_n P_j)' \rangle - \langle \phi \mathbf{u}, P'_n P'_j \rangle + \langle \psi \mathbf{u}, P'_n P_j \rangle = -\langle \phi \mathbf{u}, P'_n P'_j \rangle.$$

Since by hypothesis C1 holds, and we have already proved that C1 ⇒ C1′ ⇒ C2 ⇒ C3, and in the proof of C2 ⇒ C3 we have shown that $\{Q_n := P'_{n+1}/(n+1)\}_{n \geq 0}$ is a monic OPS with respect to $\mathbf{v} := \phi \mathbf{u}$, then $\langle \phi \mathbf{u}, P'_n P'_j \rangle = 0$ if $j \neq n$, hence (8.22) reduces to

$$\phi P''_n + \psi P'_n + \lambda_n P_n = 0, \tag{8.23}$$

where $\lambda_n := -\lambda_{n,n}$. Comparing leading coefficients in (8.23), and setting $\phi(x) = ax^2 + bx + c$ and $\psi(x) = px + q$, we obtain $\lambda_n = -n\left((n-1)a + p\right) =$

$-nd_{n-1}$, hence $\lambda_n \neq 0$ if $n \geq 1$ (since C1 ⇒ C1′, so (ϕ, ψ) is an admissible pair). Thus C1 ⇒ C5.

(C5 ⇒ C1). By hypothesis, there exist $\phi, \psi \in \mathcal{P}$, and $\lambda_n \in \mathbb{C}$, with $\lambda_n \neq 0$ if $n \geq 1$, such that $-\phi P''_{n+1} = \psi P'_{n+1} + \lambda_{n+1} P_{n+1}$. Taking in this equation $n = 0$ and $n = 1$ we deduce

$$\psi = -\lambda_1 P_1 \in \mathcal{P}_1 \setminus \mathcal{P}_0, \quad \phi = -(\psi P'_2 + \lambda_2 P_2)/2 \in \mathcal{P}_2.$$

We will prove that $D(\phi\mathbf{u}) = \psi\mathbf{u}$ by showing that the actions of the functionals $D(\phi\mathbf{u})$ and $\psi\mathbf{u}$ coincide on the simple set $\{Q_n\}_{n\geq 0}$. Indeed,

$$\begin{aligned}\langle D(\phi\mathbf{u}), Q_n\rangle &= \frac{1}{n+1}\langle D(\phi\mathbf{u}), P'_{n+1}\rangle \\ &= -\frac{1}{n+1}\langle \mathbf{u}, \phi P''_{n+1}\rangle \frac{1}{n+1}\langle \mathbf{u}, \psi P'_{n+1} + \lambda_{n+1} P_{n+1}\rangle \\ &= \langle \mathbf{u}, \psi Q_n\rangle + \frac{\lambda_{n+1}}{n+1}\langle \mathbf{u}, P_{n+1}\rangle = \langle \psi\mathbf{u}, Q_n\rangle.\end{aligned}$$

Since at least one of the polynomials ϕ and ψ is nonzero (because $\lambda_n \neq 0$), C1 holds.

(C2 ⇒ C7). Since by hypothesis (C2) holds, we may write

$$\phi P'_n = a_n P_{n+1} + b_n P_n + c_n P_{n-1}, \tag{8.24}$$

$$\phi P'_{n-1} = a_{n-1} P_n + b_{n-1} P_{n-1} + c_{n-1} P_{n-2}. \tag{8.25}$$

Multiplying (8.24) by P_{n-1} and (8.25) by P_n and adding the resulting equalities, we find that $\phi(P_n P_{n-1})'$ is a linear combination of the polynomials P_n^2, $P_n P_{n-1}$, P_{n-1}^2, $P_{n+1} P_{n-1}$, and $P_n P_{n-2}$. Substituting P_{n+1} and P_{n-2} by the corresponding expressions are given by the three-term recurrence relation, we deduce

$$\phi(P_n P_{n-1})' = A_n P_n^2 + (B_n x + C_n) P_n P_{n-1} + D_n P_{n-1}^2, \quad n \in \mathbb{N}, \tag{8.26}$$

where

$$A_n := a_{n-1} - \frac{c_{n-1}}{\gamma_{n-1}}, \qquad B_n := a_n + \frac{c_{n-1}}{\gamma_{n-1}},$$

$$C_n := -a_n\beta_n + b_n + b_{n-1} - \frac{c_{n-1}}{\gamma_{n-1}}\beta_{n-1}, \qquad D_n := c_n - a_n\gamma_n.$$

Write $\phi(x) = ax^2 + bx + c$ and $\psi(x) = px + q$. We have already seen that C2 ⇔ C1′, and while proving C1′ ⇒ C2 we have shown that the coefficients a_n, b_n, and c_n appearing in (8.24) are given by $a_n = na$, $b_n = -\frac{1}{2}\psi(\beta_n)$, and $c_n = -d_{n-1}\gamma_n$. Hence

$$A_n = d_{2n-3}, \quad B_n = 2a - p, \quad D_n = -d_{2n-1}\gamma_n, \tag{8.27}$$

$$C_n = -\frac{1}{2}\left(d_{2n}\beta_n - d_{2n-4}\beta_{n-1}\right) - q = b - q, \tag{8.28}$$

where the last equality is easily derived using the expressions for the $\beta-$parameters given in the statement of the theorem. Therefore, $B_n x + C_n = \phi' - \psi$ (independent of n). Finally, substituting (8.27) and (8.28) into (8.26) yields the equation appearing in C7, being $h_n = A_n = d_{2n-3}$ and $t_n = D_n = -d_{2n-1}\gamma_n$, and so C2 ⇒ C7.

(C7 ⇒ C2). Fix an integer $n \in \mathbb{N}$. For this n, rewrite the equation in (C7) as

$$(\phi P_n' + \psi P_n - t_n P_{n-1})\, P_{n-1} = \left(-\phi P_{n-1}' + \phi' P_{n-1} + h_n P_n\right) P_n.$$

Therefore, since P_n and P_{n-1} have no common zeros, there is $\pi_{1,n} \in \mathcal{P}_1$ such that

$$\phi P_n' + \psi P_n - t_n P_{n-1} = \pi_{1,n} P_n, \tag{8.29}$$

$$-\phi P_{n-1}' + \phi' P_{n-1} + h_n P_n = \pi_{1,n} P_{n-1}. \tag{8.30}$$

By comparing the leading coefficients on both sides of equation (8.29) we deduce $\pi_{1,n}(x) = d_n x + z_n$ for some $z_n \in \mathbb{C}$ (and $d_n := na + p$). By hypothesis, (ϕ, ψ) is an admissible pair, hence $d_n \neq 0$ and so $\deg \pi_{1,n} = 1$. Moreover, by the three-term recurrence relation for $\{P_n\}_{n\geq 0}$, $xP_n = P_{n+1} + \beta_n P_n + \gamma_n P_{n-1}$. Therefore, (8.29) may be rewritten as

$$\phi P_n' = a_n P_{n+1} + b_n P_n + c_n P_{n-1},$$

where $a_n := na$, $b_n := na\beta_n + z_n - q$, and $c_n := na\gamma_n + t_n$. To conclude the proof we need to show that $c_n \neq 0$ for all $n \in \mathbb{N}$. Indeed, changing n into $n+1$ in (8.30) and adding the resulting equation with (8.29), we obtain

$$\begin{aligned}(\psi(x) + \phi'(x))P_n(x) - t_n P_{n-1}(x) + h_{n+1}P_{n+1}(x)\\ = ((d_n + d_{n+1})x + (z_n + z_{n+1}))\, P_n(x).\end{aligned}$$

Since $\psi(x) + \phi'(x) = (2a + p)x + q + b$ and taking into account once again the three-term recurrence relation for $\{P_n\}_{n\geq 0}$, the last equation may be rewritten as a trivial linear combination of the three polynomials P_{n+1}, P_n, and P_{n-1}. Thus, we deduce

$$h_{n+1} = d_{2n-1}, \quad z_{n+1} = -z_n - d_{2n-1}\beta_n + q + b, \quad t_n = -d_{2n-1}\gamma_n.$$

Therefore, $c_n = na\gamma_n + t_n = -d_{n-1}\gamma_n \neq 0$. This completes the proof. □

Remark 8.2. *The β and γ-parameters in Theorem 8.1 may be written explicitly in terms (only) of the coefficients of ϕ and ψ as follows:*

$$\beta_n = -\frac{(-2a + p)q + 2bn((n-1)a + p)}{(2na + p)[(2n-2)a + p]},$$

$$\gamma_{n+1} = -(n+1)$$
$$\times \frac{((n-1)a+p)(a(nb+q)^2 - b(nb+q)(2na+p) + c(2na+p)^2)}{((2n-1)a+p)(2na+p)^2(2(n+1)a+p)}.$$

It is worth mentioning that the functional approach considered here simplifies dramatically the original proofs of the characterizations in the Theorem 8.1. The reader is invited to look at some of the original proofs; see, for instance, [1, 7, 21, 22, 41]. The statement and proof of Theorem 8.1 are based on [35, 36, 30, 11]. C4 and C4′ can be found in [11]. However, C4 for $k=1$ was proved by Geronimus [19]. The proof of C4$\Rightarrow$C1 uses arguments originally presented in [38].

8.2 Classification and Canonical Representatives

In this section we prove a remarkable property: *Up to constant factors and affine changes of variables, there are only four (parametric) families of classical OP, namely, Hermite, Laguerre, Jacobi, and Bessel polynomials.* The corresponding regular functionals will be denoted by $\mathbf{u}_H$, $\mathbf{u}_L^{(\alpha)}$, $\mathbf{u}_J^{(\alpha,\beta)}$, and $\mathbf{u}_B^{(\alpha)}$, respectively, and these are called the canonical representatives (or canonical forms) of the classical functionals. Their description is given in Table 8.1. Each one of these functionals fulfills the functional differential equation (8.1), being the corresponding pair $(\phi,\psi)\equiv(\Phi,\Psi)$ given in the table. The regularity conditions in the table are determined by conditions (ii) appearing in (8.4).

Ultimately, denoting by $[\mathbf{u}]$ the equivalent class determined by a functional $\mathbf{u}\in\mathcal{P}'$, and setting $\mathcal{P}'_C := \{\mathbf{u}\in\mathcal{P}' \,|\, \mathbf{u} \text{ is classical}\}$, we will show that

$$\mathcal{P}'_C/\!\sim\; := \{\,[\mathbf{u}]\,|\,\mathbf{u}\in\mathcal{P}'_C\} = \Big\{[\mathbf{u}_H], [\mathbf{u}_L^{(\alpha)}], [\mathbf{u}_J^{(\alpha,\beta)}], [\mathbf{u}_B^{(\alpha)}]\Big\},$$

where the parameters α and β vary on $\mathbb{C}$ subject to the regularity conditions in Table 8.1, and $\sim$ is the equivalence relation in $\mathcal{P}'$ introduced in Theorem 6.2, defined by

$$\mathbf{u}\sim\mathbf{v} \qquad \text{iff} \qquad \exists\, A\in\mathbb{C}\setminus\{0\},\ \exists\, B\in\mathbb{C}\ :\ \mathbf{v} = (\boldsymbol{h}_{A^{-1}}\circ\boldsymbol{\tau}_{-B})\,\mathbf{u}. \tag{8.31}$$

We start by proving a result that allows us to ensure that this equivalence relation preserves the classical character of a given classical functional.

TABLE 8.1: Canonical forms.

$\mathbf{u}$	Φ	Ψ	Regularity Conditions
$\mathbf{u}_H$	1	$-2x$	——
$\mathbf{u}_L^{(\alpha)}$	x	$-x+\alpha+1$	$-\alpha \notin \mathbb{N}$
$\mathbf{u}_J^{(\alpha,\beta)}$	$1-x^2$	$-(\alpha+\beta+2)x+\beta-\alpha$	$-\alpha, -\beta, -(\alpha+\beta+1) \notin \mathbb{N}$
$\mathbf{u}_B^{(\alpha)}$	x^2	$(\alpha+2)x+2$	$-(\alpha+1) \notin \mathbb{N}$

Lemma 8.1. *Let* $\mathbf{u}, \mathbf{v} \in \mathcal{P}'$ *and suppose that* $\mathbf{u} \sim \mathbf{v}$, *i.e.*, (8.31) *holds. Suppose that there exist two polynomials* ϕ *and* ψ *such that*

$$D(\phi \mathbf{u}) = \psi \mathbf{u}.$$

Let $\Phi(x) := K\phi(Ax+B)$ *and* $\Psi(x) := KA\psi(Ax+B)$, $B \in \mathbb{C}$, $A, K \in \mathbb{C} \setminus \{0\}$. *Then*

$$D(\Phi \mathbf{v}) = \Psi \mathbf{v}.$$

Moreover, if $\mathbf{u}$ *is a classical functional, then so is* $\mathbf{v}$.

Proof. Since $\mathbf{u}$ and $\mathbf{v}$ fulfill (8.31), we have

$$\langle \mathbf{v}, x^n \rangle = \left\langle \mathbf{u}, \left(\frac{x-B}{A} \right)^n \right\rangle, \quad n \in \mathbb{N}_0.$$

Hence

$$\begin{aligned}
\langle D(\Phi \mathbf{v}), x^n \rangle &= -n \langle \mathbf{v}, \Phi(x) x^{n-1} \rangle = -n \left\langle \mathbf{u}, (\tau_B \circ h_{A^{-1}}) \left(\Phi(x) x^{n-1} \right) \right\rangle \\
&= -n \left\langle \mathbf{u}, \Phi\left(\frac{x-B}{A} \right) \left(\frac{x-B}{A} \right)^{n-1} \right\rangle \\
&= -\left\langle \mathbf{u}, K\phi(x) \cdot A \frac{\mathrm{d}}{\mathrm{d}x} \left(\left(\frac{x-B}{A} \right)^n \right) \right\rangle \\
&= KA \left\langle D\left(\phi(x)\mathbf{u}\right), \left(\frac{x-B}{A} \right)^n \right\rangle = KA \left\langle \psi(x)\mathbf{u}, \left(\frac{x-B}{A} \right)^n \right\rangle \\
&= \left\langle \mathbf{u}, \Psi\left(\frac{x-B}{A} \right) \left(\frac{x-B}{A} \right)^n \right\rangle = \left\langle \mathbf{u}, (\tau_B \circ h_{A^{-1}}) \left(\Psi(x) x^n \right) \right\rangle \\
&= \langle \mathbf{v}, \Psi(x) x^n \rangle = \langle \Psi \mathbf{v}, x^n \rangle.
\end{aligned}$$

Finally, the last sentence stated in the lemma follows from Theorem 6.4. □

Theorem 8.2 (Canonical Representatives). *Let $\mathbf{u} \in \mathcal{P}'$ be a classical functional such that*

$$D(\phi\mathbf{u}) = \psi\mathbf{u}, \tag{8.32}$$

where $\phi(x) := ax^2 + bx + c$ and $\psi(x) := px + q$, subject to the regularity conditions

$$na + p \neq 0, \quad \phi\left(-\frac{nb+q}{2na+p}\right) \neq 0, \quad n \in \mathbb{N}_0. \tag{8.33}$$

Then, there exists a regular functional $\mathbf{v} \in \mathcal{P}'$ such that

$$\mathbf{u} \sim \mathbf{v}, \quad D(\Phi\mathbf{v}) = \Psi\mathbf{v}, \tag{8.34}$$

where, for each classical functional determined by the pair (ϕ, ψ), the corresponding pair (Φ, Ψ) is given in Table 8.1. More precisely, setting

$$\Delta := b^2 - 4ac; \quad d := \psi\left(-\frac{b}{2a}\right) \quad \text{if } a \neq 0,$$

the following hold:

1. (Hermite) *If $a = b = 0$,*
$$\mathbf{v} = \left(\boldsymbol{h}_{\sqrt{-p/(2c)}} \circ \boldsymbol{\tau}_{q/p}\right)\mathbf{u} = \mathbf{u}_{_H};$$
2. (Laguerre) *if $a = 0$ and $b \neq 0$,*
$$\mathbf{v} = \left(\boldsymbol{h}_{-p/b} \circ \boldsymbol{\tau}_{c/b}\right)\mathbf{u} = \mathbf{u}_{_L}^{(\alpha)}, \quad \alpha := -1 + (qb - pc)/b^2;$$
3. (Bessel) *if $a \neq 0$ and $\Delta = 0$,*
$$\mathbf{v} = \left(\boldsymbol{h}_{2a/d} \circ \boldsymbol{\tau}_{b/(2a)}\right)\mathbf{u} = \mathbf{u}_{_B}^{(\alpha)}, \quad \alpha := -2 + p/a;$$
4. (Jacobi) *if $a \neq 0$ and $\Delta \neq 0$,*
$$\mathbf{v} = \left(\boldsymbol{h}_{-2a/\sqrt{\Delta}} \circ \boldsymbol{\tau}_{b/(2a)}\right)\mathbf{u} = \mathbf{u}_{_J}^{(\alpha,\beta)},$$
$$\alpha := -1 + p/(2a) - d/\sqrt{\Delta}, \quad \beta := -1 + p/(2a) + d/\sqrt{\Delta}.$$

Proof. Taking into account Lemma 8.1, the theorem will be proved if we are able to show that, for each given pair (ϕ, ψ), and for each corresponding pair

(Φ, Ψ) given by Table 8.1—where the "corresponding pair" (Φ, Ψ) is the one in the table such that ϕ and Φ have the same degree and their zeros have the same multiplicity—, there exist $A, K \in \mathbb{C} \setminus \{0\}$ and $B \in \mathbb{C}$ such that the relations

$$\Phi(x) = K\phi(Ax+B), \quad \Psi(x) = KA\psi(Ax+B) = KA^2px+KA(Bp+q) \quad (8.35)$$

hold, for appropriate choices of the parameters α and β appearing in Table 8.1 for the Laguerre, Bessel, and Jacobi cases. Indeed, considering the four possible cases determined by the polynomial ϕ, we have:

1. Assume $a = b = 0$, i.e., $\phi(x) = c$. The regularity conditions (8.33) ensure that $p \neq 0$ and $c \neq 0$. Therefore, since in this case we require $(\Phi, \Psi) = (1, -2x)$, from (8.35) we obtain the equations

$$1 = Kc, \quad -2 = KA^2p, \quad 0 = Bp + q.$$

A solution of this system of equations is

$$K = 1/c, \quad A = \sqrt{-2c/p}, \quad B = -q/p,$$

which gives, by Lemma 8.1, the desired result for the Hermite case.

2. Assume $a = 0$ and $b \neq 0$, so that $\phi(x) = bx + c$. Since in this case we require $(\Phi, \Psi) = (x, -x + \alpha + 1)$, from (8.35) we obtain

$$1 = KAb, \quad 0 = bB + c, \quad -1 = KA^2p, \quad \alpha + 1 = KA(Bp + q).$$

Solving this system we find

$$K = -p/b^2, \quad B = -c/b, \quad A = -b/p, \quad \alpha = -1 + (qb - pc)/b^2.$$

Notice that, in this case,

$$d_n = p, \quad \phi\left(-\frac{nb + q}{2na + p}\right) = -\frac{b^2}{p}(n + \alpha + 1),$$

hence the regularity conditions (8.33) ensure that $p \neq 0$ (and so K and A are well defined, being both nonzero complex numbers) and $-\alpha \notin \mathbb{N}$.

3. Assume $a \neq 0$ and $\Delta = 0$. Then $\phi(x) = a\left(x + \frac{b}{2a}\right)^2$. In this case we require $(\Phi, \Psi) = \left(x^2, (\alpha + 2)x + 2\right)$, hence from (8.35) we obtain

$$1 = KA^2a, \quad 0 = B + b/(2a), \quad \alpha + 2 = KA^2p, \quad 2 = KA(Bp + q).$$

Therefore, taking into account that $d := \psi\left(-\frac{b}{2a}\right) = (2aq-pb)/(2a)$, we deduce

$$K = 4a/d^2, \quad B = -b/(2a), \quad A = d/(2a), \quad \alpha = -2 + p/a.$$

In this case we have

$$d_n = a(n + \alpha + 2), \quad \phi\left(-\frac{nb + q}{2na + p}\right) = \frac{d^2}{a(2n + \alpha + 2)^2},$$

hence conditions (8.33) ensure that $-(\alpha+1)\notin\mathbb{N}$ and $d\neq 0$, and so, in particular, K is well defined, being both K and A nonzero complex numbers.

4. Finally, assume $a\neq 0$ and $\Delta\neq 0$. Writing

$$\phi(x)=a\left(\left(x+\frac{b}{2a}\right)^2-\frac{\Delta}{4a^2}\right),$$

since in this case we require $(\Phi,\Psi)=\big(1-x^2,-(\alpha+\beta+2)x+\beta-\alpha\big)$, from (8.35) we obtain

$$-1=KA^2a,\qquad B+b/(2a)=0,$$

$$1=Ka\left(\left(B+\frac{b}{2a}\right)^2-\frac{\Delta}{4a^2}\right),\qquad -(\alpha+\beta+2)=KA^2p,$$

$$\beta-\alpha=KA(Bp+q).$$

A solution of this system of five equations is

$$A=-\sqrt{\Delta}/(2a),\qquad B=-b/(2a),\qquad K=-4a/\Delta,$$

$$\alpha=-1+p/(2a)-d/\sqrt{\Delta},\qquad \beta=-1+p/(2a)+d/\sqrt{\Delta}.$$

(We choose A with the minus sign since whenever $(\phi,\psi)=(\Phi,\Psi)$ that choice implies $A=1$ and $B=0$, hence $\mathbf{u}=\mathbf{v}=\mathbf{u}_J^{(\alpha,\beta)}$, and so it is a more natural choice.) Adding and subtracting the last equations for α and β, we find $\alpha+\beta+2=p/a$ and $\alpha-\beta=-2d/\sqrt{\Delta}$, hence we deduce

$$d_n=a(n+\alpha+\beta+2),\quad \phi\left(-\frac{nb+q}{2na+p}\right)=-\frac{\Delta}{a}\frac{(n+\alpha+1)(n+\beta+1)}{(2n+\alpha+\beta+2)^2},$$

Therefore, conditions (8.33) ensure that $-(\alpha+\beta+1)\notin\mathbb{N}$, $-\alpha\notin\mathbb{N}$, and $-\beta\notin\mathbb{N}$, and the theorem is proved. □

Remark 8.3. *It follows from the proof of Theorem 8.2 that the parameters α and β (in cases 2, 3, and 4) fulfill the regularity conditions appearing in Table 8.1.*

The preceding theorem allows us to classify each classical functional according with the degree of the polynomial ϕ appearing in the functional differential equation (8.1).

Corollary 8.1. *Let* $\mathbf{u}$ *be a classical functional fulfilling* (8.1)–(8.2).

(i) *If* $\deg\phi = 0$ *(hence* ϕ *is a nonzero constant), then* $\mathbf{u} \sim \mathbf{u}_H$*;*

(ii) *if* $\deg\phi = 1$*, then* $\mathbf{u} \sim \mathbf{u}_L^{(\alpha)}$ *for some* α*;*

(iii) *if* $\deg\phi = 2$ *and* ϕ *has simple zeros, then* $\mathbf{u} \sim \mathbf{u}_J^{(\alpha,\beta)}$ *for some pair* (α, β)*;*

(iv) *if* $\deg\phi = 2$ *and* ϕ *has a double zero, then* $\mathbf{u} \sim \mathbf{u}_B^{(\alpha)}$ *for some* α*.*

Remark 8.4. *The monic OPS with respect to the canonical representatives* $\mathbf{u}_H$, $\mathbf{u}_L^{(\alpha)}$, $\mathbf{u}_J^{(\alpha,\beta)}$*, and* $\mathbf{u}_B^{(\alpha)}$ *will be denoted by* $\{\widehat{H}_n\}_{n\geq 0}$, $\{\widehat{L}_n^{(\alpha)}\}_{n\geq 0}$, $\{\widehat{P}_n^{(\alpha,\beta)}\}_{n\geq 0}$*, and* $\{\widehat{B}_n^{(\alpha)}\}_{n\geq 0}$*, respectively, and they will be called the (monic) Hermite, Laguerre, Jacobi, and Bessel polynomials. Table 8.2 and Table 8.3 summarize the corresponding parameters appearing in all characterizations presented in the Theorem 8.1.*

In view of Theorems 8.2 and 6.4, we may now justify a sentence made at the beginning of the section. As Maroni said in an interview when asked about Bessel polynomials: *"comme dans le roman d'Alexandre Dumas, les trois mousquetaires étaient quatre en réalité"*. Notice also the following special cases of Jacobi polynomials (up to normalization) that we have introduced in some previous texts: $\alpha = \beta = 0$, Legendre polynomials; $\alpha = \beta = -1/2$, Chebyshev polynomials of the first kind $\{T_n\}_{n\geq 0}$; $\alpha = \beta = 1/2$, Chebyshev polynomials of the second kind $\{U_n\}_{n\geq 0}$; $\alpha = \beta =: \lambda - 1/2$, Gegenbauer (or ultraspherical) polynomials $\{C_n^\lambda\}_{n\geq 0}$, $-2\lambda \notin \mathbb{N}$.

TABLE 8.2: Parameters appearing in Theorem 8.1.

	$\widehat{H}_n$	$\widehat{L}_n^{(\alpha)}$	$\widehat{P}_n^{(\alpha,\beta)}$	$\widehat{B}_n^{(\alpha)}$
λ_n	$2n$	n	$n(n+\alpha+\beta+1)$	$-n(n+\alpha+1)$
β_n	0	$2n+\alpha+1$	$\frac{\beta^2-\alpha^2}{(2n+\alpha+\beta)(2n+2+\alpha+\beta)}$	$\frac{-2\alpha}{(2n+\alpha)(2n+2+\alpha)}$
γ_n	$\frac{n}{2}$	$n(n+\alpha)$	$\frac{4n(n+\alpha)(n+\beta)(n+\alpha+\beta)}{(2n+\alpha+\beta-1)(2n+\alpha+\beta)^2(2n+\alpha+\beta+1)}$	$\frac{-4n(n+\alpha)}{(2n+\alpha-1)(2n+\alpha)^2(2n+\alpha+1)}$
a_n	0	0	$-n$	n
b_n	0	n	$\frac{2(\alpha-\beta)n(n+\alpha+\beta+1)}{(2n+\alpha+\beta)(2n+2+\alpha+\beta)}$	$\frac{-4n(n+\alpha+1)}{(2n+\alpha)(2n+2+\alpha)}$
c_n	n	$n(n+\alpha)$	$\frac{4n(n+\alpha)(n+\beta)(n+\alpha+\beta)(n+\alpha+\beta+1)}{(2n+\alpha+\beta-1)(2n+\alpha+\beta)^2(2n+\alpha+\beta+1)}$	$\frac{4n(n+\alpha)(n+\alpha+1)}{(2n+\alpha-1)(2n+\alpha)^2(2n+\alpha+1)}$

TABLE 8.3: Parameters appearing in Theorem 8.1.

	$\widehat{H}_n$	$\widehat{L}_n^{(\alpha)}$	$\widehat{P}_n^{(\alpha,\beta)}$	$\widehat{B}_n^{(\alpha)}$
$r_n^{[1]}$	0	n	$\frac{2(\alpha-\beta)n}{(2n+\alpha+\beta)(2n+2+\alpha+\beta)}$	$\frac{4n}{(2n+\alpha)(2n+2+\alpha)}$
$s_n^{[1]}$	0	0	$\frac{-4(n-1)n(n+\alpha)(n+\beta)}{(2n+\alpha+\beta-1)(2n+\alpha+\beta)^2(2n+\alpha+\beta+1)}$	$\frac{4(n-1)n}{(2n+\alpha-1)(2n+\alpha)^2(2n+\alpha+1)}$
h_n	-2	-1	$-(2n+\alpha+\beta-1)$	$2n+\alpha-1$
t_n	n	$n(n+\alpha)$	$\frac{4n(n+\alpha)(n+\beta)(n+\alpha+\beta)}{(2n+\alpha+\beta-1)(2n+\alpha+\beta)^2}$	$\frac{4n(n+\alpha)}{(2n+\alpha-1)(2n+\alpha)^2}$
k_n	$\frac{(-1)^n}{2^n}$	$(-1)^n$	$\frac{(-1)^n}{(n+\alpha+\beta+1)_n}$	$\frac{1}{(n+\alpha+1)_n}$

Exercises

1. Complete the proof of Theorem 8.1 by proving that:

   (a) The polynomial Δ_2 introduced in the proof of C4 $\Rightarrow$ C1 fulfills $\Delta_2 \in \mathcal{P}_2 \setminus \{0\}$.

   (b) C1$'$ $\Leftrightarrow$ C6.

2. For arbitrary $k \in \mathbb{N}$, find expressions for the parameters $r_n^{[k]}$ and $s_n^{[k]}$ appearing in characterization C4 of Theorem 8.1, only in terms of the coefficients of the polynomials ϕ and ψ appearing in the functional differential equation for $\mathbf{u}$. Compute these expressions for the classical canonical forms of Hermite, Laguerre, Jacobi, and Bessel.

3. Let $\mathbf{u}$ be a classical functional so that it is a regular functional on $\mathcal{P}$ which fulfills the distributional differential equation $D(\phi\mathbf{u}) = \psi\mathbf{u}$, being $\phi \in \mathcal{P}_2$ and $\psi \in \mathcal{P}_1 \setminus \mathcal{P}_0$.

   (a) Find a closed formula for the Hankel determinant $H_n := \det\left([u_{i+j}]_{i,j=0}^{n}\right)$ (of order $n+1$), involving only the (coefficients of the) polynomials ϕ and ψ.

   (b) Compute H_n for the classical canonical forms of Hermite, Laguerre, Jacobi, and Bessel. Give also expressions for the moments in each case.

Chapter 9

Functional Equation on Lattices

In this chapter we present necessary and sufficient conditions for the regularity of solutions of the functional equation $\mathbf{D}(\phi\mathbf{u}) = \mathbf{S}(\psi\mathbf{u})$ on lattices, where $\phi \in \mathcal{P}_2$ and $\psi \in \mathcal{P}_1$. We also present the functional Rodrigues formula and a closed formula for the recurrence coefficients. The main source of this chapter is the article [10] (see also [39]).

9.1 Functional Equation on Lattices

In view of Chapter 7, a natural question arises related to the analogs of Theorem 7.2 on lattices. The aim of this chapter is to answer this question. Let $x(s)$ be a lattice given by (1.9). The functional differential equation considered in this chapter has the form

$$\mathbf{D}(\phi\mathbf{u}) = \mathbf{S}(\psi\mathbf{u}). \tag{9.1}$$

where $\phi \in \mathcal{P}_2$ and $\psi \in \mathcal{P}_1$, and $\mathbf{u} \in \mathcal{P}'$ is the unknown. Recall that $\mathbf{D}$ and $\mathbf{S}$, as well as D and S, introduced in Definition 1.3, depend on the lattice. Notice also that, as in the previous chapter, we do not require *a priori* $\mathbf{u}$ to be a regular functional. The next definition appears in [17], and it extends Definition 8.1.

Definition 9.1. *Let $x(s)$ be a lattice given by (1.9). The functional $\mathbf{u} \in \mathcal{P}'$ is called a x-classical functional if the following two conditions hold:*

(i) $\mathbf{u}$ *is regular;*

(ii) $\mathbf{u}$ *satisfies the functional differential equation*

$$\mathbf{D}(\phi\mathbf{u}) = \mathbf{S}(\psi\mathbf{u}), \tag{9.2}$$

where ϕ and ψ are polynomials fulfilling

$$\deg\phi \leq 2, \quad \deg\psi = 1.$$

DOI: 10.1201/9781003432067-9

An OPS $\{P_n\}_{n\geq 0}$ with respect to a x-classical functional is called a x-classical OPS.

We will refer to (9.2) as a functional equation on lattices. As mentioned above, our principal goal in this chapter is to state the necessary and sufficient conditions, involving only ϕ and ψ (or, equivalently, their coefficients), such that a given functional $\mathbf{u} \in \mathcal{P}'$ satisfying the functional equation (9.2) becomes regular. In order to move on we need to introduce some notation and prove some preliminary properties. In this chapter we denote by $R_n^{[k]}$ the monic polynomial of degree n defined by

$$R_n^{[k]}(z) := \frac{\mathrm{D}^k R_{n+k}(z)}{\prod_{j=1}^{k} \gamma_{n+j}} = \frac{\gamma_n!}{\gamma_{n+k}!}\mathrm{D}^k R_{n+k}(z), \quad k, n \in \mathbb{N}_0. \tag{9.3}$$

Here, as usual, it is understood that $\mathrm{D}^0 f = f$, empty product equals one. The next definition extends Definition 8.2.

Definition 9.2. *Let $\phi \in \mathcal{P}_2$ and $\psi \in \mathcal{P}_1$. (ϕ, ψ) is called an x-admissible pair if*

$$d_n \equiv d_n(\phi, \psi, x) := \frac{1}{2}\gamma_n \phi'' + \alpha_n \psi' \neq 0, \quad n \in \mathbb{N}_0.$$

9.1.1 Preliminary Properties

Given $\mathbf{u} \in \mathcal{P}'$, $\phi \in \mathcal{P}_2$, and $\psi \in \mathcal{P}_1$, we define recursively polynomials $\phi^{[k]} \in \mathcal{P}_2$ and $\psi^{[k]} \in \mathcal{P}_1$ by

$$\phi^{[0]} := \phi, \quad \psi^{[0]} := \psi, \tag{9.4}$$

$$\phi^{[k+1]} := \mathrm{S}\phi^{[k]} + \mathrm{U}_1\mathrm{S}\psi^{[k]} + \alpha\mathrm{U}_2\mathrm{D}\psi^{[k]}, \tag{9.5}$$

$$\psi^{[k+1]} := \mathrm{D}\phi^{[k]} + \alpha\mathrm{S}\psi^{[k]} + \mathrm{U}_1\mathrm{D}\psi^{[k]}, \tag{9.6}$$

and functionals $\mathbf{u}^{[k]} \in \mathcal{P}'$ by

$$\mathbf{u}^{[0]} := \mathbf{u}, \quad \mathbf{u}^{[k+1]} := \mathbf{D}\left(\mathrm{U}_2\psi^{[k]}\mathbf{u}^{[k]}\right) - \mathbf{S}\left(\phi^{[k]}\mathbf{u}^{[k]}\right), \quad k \in \mathbb{N}_0. \tag{9.7}$$

The functional $\mathbf{u}^{[k]}$ may be seen as the highest-order x-derivative of $\mathbf{u}$. Next, we provide explicit representations for the polynomials $\phi^{[k]}$ and $\psi^{[k]}$.

Proposition 9.1. *For the lattice $x(s) = \mathfrak{c}_1 q^{-s} + \mathfrak{c}_2 q^s + \mathfrak{c}_3$, the following identities hold:*

$$\mathrm{D}z^n = \gamma_n z^{n-1} + u_n z^{n-2} + v_n z^{n-3} + \cdots, \quad n \in \mathbb{N}_0, \tag{9.8}$$

$$\mathrm{S}z^n = \alpha_n z^n + \widehat{u}_n z^{n-1} + \widehat{v}_n z^{n-2} + \cdots, \quad n \in \mathbb{N}_0, \tag{9.9}$$

where α_n and γ_n are given in Theorem 1.4, and

$$u_n := \big(n\gamma_{n-1} - (n-1)\gamma_n\big)\mathfrak{c}_3, \tag{9.10}$$

$$v_n := \big(n\gamma_{n-2} - (n-2)\gamma_n\big)\mathfrak{c}_1\mathfrak{c}_2 \tag{9.11}$$
$$+ \frac{1}{2}\big(n(n-1)\gamma_{n-2} - 2n(n-2)\gamma_{n-1} + (n-1)(n-2)\gamma_n\big)\mathfrak{c}_3^2,$$

$$\widehat{u}_n := n(\alpha_{n-1} - \alpha_n)\mathfrak{c}_3, \tag{9.12}$$

$$\widehat{v}_n := n(\alpha_{n-2} - \alpha_n)\mathfrak{c}_1\mathfrak{c}_2 + n(n-1)(\alpha-1)\alpha_{n-1}\mathfrak{c}_3^2. \tag{9.13}$$

Proof. The proof is by induction on n. For $n = 0$, we have $\mathrm{D}z^0 = \mathrm{D}1 = 0$ and $\mathrm{S}z^0 = 1$. Since $\gamma_0 = 0$ and $\alpha_0 = 1$, we see that (9.8)–(9.9) hold for $n = 0$. Next, suppose that relations (9.8)–(9.9) are true for all integer numbers less than or equal to a fixed nonnegative integer number n (induction hypothesis). Using this hypothesis together with (1.24) and (1.43), we obtain

$$\begin{aligned}\mathrm{D}z^{n+1} &= \mathrm{D}z^n\mathrm{S}z + \mathrm{S}z^n\mathrm{D}z = (\alpha z + \beta)\mathrm{D}z^n + \mathrm{S}z^n \\ &= (\alpha_n + \alpha\gamma_n)z^n + (\alpha u_n + \widehat{u}_n + \beta\gamma_n)z^{n-1} \\ &\quad + (\alpha v_n + \widehat{v}_n + \beta u_n)z^{n-2} + \cdots .\end{aligned}$$

Similarly,

$$\begin{aligned}\mathrm{S}z^{n+1} &= \mathtt{U}_2(z)\mathrm{D}z\;\mathrm{D}z^n + \mathrm{S}z^n\;\mathrm{S}z = \mathtt{U}_2(z)\mathrm{D}z^n + (\alpha z + \beta)\mathrm{S}z^n \\ &= \big(\alpha\alpha_n + (\alpha^2-1)\gamma_n\big)\, z^{n+1} + \big((\alpha^2-1)\,(u_n - 2\gamma_n\mathfrak{c}_3) + \alpha\widehat{u}_n + \beta\alpha_n\big)\, z^n \\ &\quad + \big((\alpha^2-1)\,\big(v_n - 2u_n\mathfrak{c}_3 + (\mathfrak{c}_3^2 - 4\mathfrak{c}_1\mathfrak{c}_2)\gamma_n\big) + \alpha\widehat{v}_n + \beta\widehat{u}_n\big)\, z^{n-1} + \cdots .\end{aligned}$$

Therefore, using relations (1.11)–(1.18), we obtain (9.8)–(9.9) with n replaced by $n+1$. Consequently, (9.8)–(9.9) hold for each nonnegative integer n, and the result follows. □

Proposition 9.2. *Consider the lattice* $x(s) := \mathfrak{c}_1 q^{-s} + \mathfrak{c}_2 q^s + \mathfrak{c}_3$. *Let* $\phi \in \mathcal{P}_2$ *and* $\psi \in \mathcal{P}_1$, *so there are* $a, b, c, d, e \in \mathbb{C}$ *such that*

$$\phi(z) = az^2 + bz + c, \quad \psi(z) = dz + e.$$

Then the polynomials $\phi^{[k]}$ *and* $\psi^{[k]}$, $k \in \mathbb{N}_0$, *defined by* (9.4)–(9.6) *are given by*

$$\psi^{[k]}(z) = (a\gamma_{2k} + d\alpha_{2k})(z - \mathfrak{c}_3) + \phi'(\mathfrak{c}_3)\gamma_k + \psi(\mathfrak{c}_3)\alpha_k, \tag{9.14}$$

$$\begin{aligned}\phi^{[k]}(z) &= \left(d(\alpha^2 - 1)\gamma_{2k} + a\alpha_{2k}\right)\left((z - \mathfrak{c}_3)^2 - 2\mathfrak{c}_1\mathfrak{c}_2\right) \\ &\quad + \left(\phi'(\mathfrak{c}_3)\alpha_k + \psi(\mathfrak{c}_3)(\alpha^2 - 1)\gamma_k\right)(z - \mathfrak{c}_3) + \phi(\mathfrak{c}_3) + 2a\mathfrak{c}_1\mathfrak{c}_2.\end{aligned} \tag{9.15}$$

Proof. Set

$$\phi^{[k]}(z) = a^{[k]}z^2 + b^{[k]}z + c^{[k]}, \quad \psi^{[k]}(z) = d^{[k]}z + e^{[k]}, \tag{9.16}$$

where $a^{[k]}, b^{[k]}, c^{[k]}, d^{[k]}, e^{[k]} \in \mathbb{C}$. Clearly, by (9.4),

$$a^{[0]} = a, \quad b^{[0]} = b, \quad c^{[0]} = c, \quad d^{[0]} = d, \quad e^{[0]} = e.$$

In order to determine the coefficients $a^{[k]}$, $b^{[k]}$, $c^{[k]}$, $d^{[k]}$, and $e^{[k]}$, we proceed as follows. Firstly we replace in (9.5) and in (9.6) the expressions of $\phi^{[k]}$, $\phi^{[k+1]}$, $\psi^{[k]}$, and $\psi^{[k+1]}$ given by (9.16); and then, in the two resulting identities, using (9.8) together with (1.19) and (1.20), after identification of the coefficients of the polynomials appearing on both sides of each of those identities, we obtain a system with five difference equations, namely

$$a^{[k+1]} = (2\alpha^2 - 1)a^{[k]} + 2\alpha(\alpha^2 - 1)d^{[k]}, \tag{9.17}$$

$$b^{[k+1]} = \alpha b^{[k]} + (\alpha^2 - 1)e^{[k]} + 2\beta(2\alpha + 1)a^{[k]} + \beta(\alpha + 1)(4\alpha - 1)d^{[k]}, \tag{9.18}$$

$$c^{[k+1]} = c^{[k]} + \widehat{v}_2 a^{[k]} + \beta b^{[k]} + \beta(\alpha + 1)e^{[k]} + \left(\beta^2(\alpha + 1) + \alpha\delta\right)d^{[k]}, \tag{9.19}$$

$$d^{[k+1]} = 2\alpha a^{[k]} + (2\alpha^2 - 1)d^{[k]}, \tag{9.20}$$

$$e^{[k+1]} = b^{[k]} + \alpha e^{[k]} + 2\beta a^{[k]} + \beta(2\alpha + 1)d^{[k]}. \tag{9.21}$$

The explicit solution of this system is

$$a^{[k]} = d(\alpha^2 - 1)\gamma_{2k} + a\alpha_{2k}, \tag{9.22}$$

$$b^{[k]} = \psi(\mathfrak{c}_3)(\alpha^2 - 1)\gamma_k + \phi'(\mathfrak{c}_3)\alpha_k - 2\mathfrak{c}_3\left(d(\alpha^2 - 1)\gamma_{2k} + a\alpha_{2k}\right), \tag{9.23}$$

$$c^{[k]} = \phi(\mathfrak{c}_3) + 2a\mathfrak{c}_1\mathfrak{c}_2 - \mathfrak{c}_3\left(\psi(\mathfrak{c}_3)(\alpha^2-1)\gamma_k + \phi'(\mathfrak{c}_3)\alpha_k\right) \quad (9.24)$$

$$+ \left(\mathfrak{c}_3^2 - 2\mathfrak{c}_1\mathfrak{c}_2\right)\left(d(\alpha^2-1)\gamma_{2k} + a\alpha_{2k}\right),$$

$$d^{[k]} = a\gamma_{2k} + d\alpha_{2k}, \quad (9.25)$$

$$e^{[k]} = \phi'(\mathfrak{c}_3)\gamma_k + \psi(\mathfrak{c}_3)\alpha_k - \mathfrak{c}_3\left(a\gamma_{2k} + d\alpha_{2k}\right). \quad (9.26)$$

This can be easily proved by induction on k. The representations (9.14)–(9.15) are obtained by inserting (9.22)–(9.26) into (9.16). □

Lemma 9.1 below is proved in [17]. We point out that the statement of the result given in [17] assumes *a priori* that $\mathbf{u}$ is a regular functional. However, an inspection of the proof given therein shows that the result remains unchanged without such an assumption.

Lemma 9.1. *Let $\mathbf{u} \in \mathcal{P}'$. Suppose that there exist $\phi \in \mathcal{P}_2$ and $\psi \in \mathcal{P}_1$ such that* (9.2) *holds. Then $\mathbf{u}^{[k]}$ satisfies the functional equation*

$$\mathbf{D}\left(\phi^{[k]}\mathbf{u}^{[k]}\right) = \mathbf{S}\left(\psi^{[k]}\mathbf{u}^{[k]}\right), \quad k \in \mathbb{N}_0. \quad (9.27)$$

The next result gives some additional functional equations fulfilled by $\mathbf{u}^{[k]}$.

Lemma 9.2. *Let $\mathbf{u} \in \mathcal{P}'$ and suppose that there exist $\phi \in \mathcal{P}_2$ and $\psi \in \mathcal{P}_1$ such that $\mathbf{u}$ satisfies the functional equation* (9.2)*. Then, for all $k \in \mathbb{N}_0$, we have*

$$\mathbf{D}\left(\mathbf{u}^{[k+1]}\right) = -\alpha\psi^{[k]}\mathbf{u}^{[k]}, \quad (9.28)$$

$$\mathbf{S}\left(\mathbf{u}^{[k+1]}\right) = -\alpha\left(\alpha\phi^{[k]} + \mathrm{U}_1\psi^{[k]}\right)\mathbf{u}^{[k]}, \quad (9.29)$$

$$2\mathrm{U}_1\mathbf{u}^{[k+1]} = \mathbf{S}\left(\mathrm{U}_2\psi^{[k]}\mathbf{u}^{[k]}\right) - \mathbf{D}\left(\mathrm{U}_2\phi^{[k]}\mathbf{u}^{[k]}\right). \quad (9.30)$$

Proof. Using (1.30) and (9.27), we deduce

$$\mathbf{D}^2\left(\mathrm{U}_2\psi^{[k]}\mathbf{u}^{[k]}\right) = (2\alpha-\alpha^{-1})\mathbf{S}^2\left(\psi^{[k]}\mathbf{u}^{[k]}\right) + \alpha^{-1}\mathrm{U}_1\mathbf{DS}\left(\psi^{[k]}\mathbf{u}^{[k]}\right) - \alpha\psi^{[k]}\mathbf{u}^{[k]}$$

$$= (2\alpha-\alpha^{-1})\mathbf{SD}\left(\phi^{[k]}\mathbf{u}^{[k]}\right) + \alpha^{-1}\mathrm{U}_1\mathbf{D}^2\left(\phi^{[k]}\mathbf{u}^{[k]}\right) - \alpha\psi^{[k]}\mathbf{u}^{[k]}$$

$$= \mathbf{DS}\left(\phi^{[k]}\mathbf{u}^{[k]}\right) - \alpha\psi^{[k]}\mathbf{u}^{[k]},$$

where the last equality follows from (1.35) for $n = 1$ and taking into account that $\alpha_2 = 2\alpha^2 - 1$ and $\gamma_1 = 1$. Therefore, by the definition of $\mathbf{u}^{[k+1]}$, we obtain

$$\mathbf{D}\mathbf{u}^{[k+1]} = \mathbf{D}^2\left(\mathtt{U}_2\psi^{[k]}\mathbf{u}^{[k]}\right), \quad -\mathbf{D}\mathbf{S}\left(\phi^{[k]}\mathbf{u}^{[k]}\right) = -\alpha\psi^{[k]}\mathbf{u}^{[k]},$$

and (9.28) follows. Next, by (1.26) and (1.27), we may write

$$\mathbf{D}\left(\mathtt{U}_2\phi^{[k]}\mathbf{u}^{[k]}\right) = \left(\mathtt{S}\mathtt{U}_2 - \alpha^{-1}\mathtt{U}_1\mathtt{D}\mathtt{U}_2\right)\mathbf{D}\left(\phi^{[k]}\mathbf{u}^{[k]}\right) + \alpha^{-1}\left(\mathtt{D}\mathtt{U}_2\right)\mathbf{S}\left(\phi^{[k]}\mathbf{u}^{[k]}\right),$$

$$\mathbf{S}\left(\mathtt{U}_2\psi^{[k]}\mathbf{u}^{[k]}\right) = \left(\mathtt{S}\mathtt{U}_2 - \alpha^{-1}\mathtt{U}_1\mathtt{D}\mathtt{U}_2\right)\mathbf{S}\left(\psi^{[k]}\mathbf{u}^{[k]}\right) + \alpha^{-1}\left(\mathtt{D}\mathtt{U}_2\right)\mathbf{D}\left(\mathtt{U}_2\psi^{[k]}\mathbf{u}^{[k]}\right).$$

After subtracting these two equalities and taking into account (9.27), as well as the relation $\alpha^{-1}\mathtt{D}\mathtt{U}_2 = 2\mathtt{U}_1$, we obtain (9.30). To prove (9.29), note first that, by the definition of $\mathbf{u}^{[k+1]}$,

$$\alpha_2\mathbf{S}\mathbf{u}^{[k+1]} = \alpha_2\mathbf{S}\mathbf{D}\left(\mathtt{U}_2\psi^{[k]}\mathbf{u}^{[k]}\right) - \alpha_2\mathbf{S}^2\left(\phi^{[k]}\mathbf{u}^{[k]}\right). \tag{9.31}$$

Using again (1.35) for $n = 1$, we have

$$\alpha_2\mathbf{S}\mathbf{D}\left(\mathtt{U}_2\psi^{[k]}\mathbf{u}^{[k]}\right) = \alpha\mathbf{D}\mathbf{S}\left(\mathtt{U}_2\psi^{[k]}\mathbf{u}^{[k]}\right) - \mathtt{U}_1\mathbf{D}^2\left(\mathtt{U}_2\psi^{[k]}\mathbf{u}^{[k]}\right)$$

and by (1.30), we also have

$$\alpha_2\mathbf{S}^2\left(\phi^{[k]}\mathbf{u}^{[k]}\right) = -\mathtt{U}_1\mathbf{D}\mathbf{S}\left(\phi^{[k]}\mathbf{u}^{[k]}\right) + \alpha^2\phi^{[k]}\mathbf{u}^{[k]} + \alpha\mathbf{D}^2\left(\mathtt{U}_2\phi^{[k]}\mathbf{u}^{[k]}\right).$$

Substituting these two expressions into the right-hand side of (9.31), we get

$$(2\alpha^2 - 1)\mathbf{S}\mathbf{u}^{[k+1]} = \alpha\mathbf{D}\mathbf{S}\left(\mathtt{U}_2\psi^{[k]}\mathbf{u}^{[k]}\right) - \alpha\mathbf{D}^2\left(\mathtt{U}_2\phi^{[k]}\mathbf{u}^{[k]}\right) - \mathtt{U}_1\mathbf{D}\mathbf{u}^{[k+1]} - \alpha^2\phi^{[k]}\mathbf{u}^{[k]}. \tag{9.32}$$

Next, by taking $f = \mathtt{U}_1$ and replacing $\mathbf{u}$ by $\mathbf{u}^{[k+1]}$ in (1.26), and then using (1.22) and (9.28), we derive

$$\mathbf{D}\left(\mathtt{U}_1\mathbf{u}^{[k+1]}\right) = -\mathtt{U}_1\psi^{[k]}\mathbf{u}^{[k]} + (\alpha - \alpha^{-1})\mathbf{S}\mathbf{u}^{[k+1]}.$$

Multiplying both sides of this equality by 2α and combining the resulting equality with the one obtained by applying $\mathbf{D}$ to both sides of (9.30), we deduce

$$(2\alpha^2 - 2)\mathbf{S}\mathbf{u}^{[k+1]} = \alpha\mathbf{D}\mathbf{S}\left(\mathtt{U}_2\psi^{[k]}\mathbf{u}^{[k]}\right) - \alpha\mathbf{D}^2\left(\mathtt{U}_2\phi^{[k]}\mathbf{u}^{[k]}\right) + 2\alpha\mathtt{U}_1\psi^{[k]}\mathbf{u}^{[k]}. \tag{9.33}$$

Finally, subtracting (9.33) from (9.32), and taking into account (9.28), (9.29) follows. □

9.1.2 The Functional Rodrigues Formula

Here we prove a functional version of the Rodrigues formula on lattices, extending the results stated in Theorem 7.1. But first, we will need to prove the following consequence of the Leibniz formula proved in Chapter 1.

Corollary 9.1. *Consider the lattice* $x(s) := \mathfrak{c}_1 q^{-s} + \mathfrak{c}_2 q^s + \mathfrak{c}_3$. *Let* $\mathbf{u} \in \mathcal{P}'$ *and* $f \in \mathcal{P}_2$. *Write* $f(z) = az^2 + bz + c$, *with* $a, b, c \in \mathbb{C}$. *Then*

$$\mathbf{D}^n(f\mathbf{u}) \tag{9.34}$$

$$= \left(\frac{a\alpha}{\alpha_n \alpha_{n-1}} (z - \mathfrak{c}_3)^2 + \frac{f'(\mathfrak{c}_3)}{\alpha_n}(z - \mathfrak{c}_3) + f(\mathfrak{c}_3) + \frac{4a(1-\alpha^2)\gamma_n \mathfrak{c}_1 \mathfrak{c}_2}{\alpha_{n-1}} \right) \mathbf{D}^n \mathbf{u}$$

$$+ \frac{\gamma_n}{\alpha_n} \left(\frac{a(\alpha_n + \alpha\alpha_{n-1})}{\alpha_{n-1}^2} (z - \mathfrak{c}_3) + f'(c_3) \right) \mathbf{D}^{n-1}\mathbf{S}\mathbf{u}$$

$$+ \frac{a\gamma_n\gamma_{n-1}}{\alpha_{n-1}^2} \mathbf{D}^{n-2}\mathbf{S}^2\mathbf{u}, \quad n \in \mathbb{N}_0. \tag{9.35}$$

In particular,

$$\mathbf{D}^n\left((bz+c)\mathbf{u}\right) = \left(\frac{b(z-\mathfrak{c}_3)}{\alpha_n} + b\mathfrak{c}_3 + c \right) \mathbf{D}^n\mathbf{u} + \frac{b\gamma_n}{\alpha_n} \mathbf{D}^{n-1}\mathbf{S}\mathbf{u}, \quad n \in \mathbb{N}_0. \tag{9.36}$$

Proof. Since (9.36) is the particular case of (9.34) for $a = 0$, we only need to prove (9.34). This can be proved by combining the Leibniz formula and identities (1.26) and (1.35). Alternatively, we may apply induction on n, as follows. Define $g(z) := f(z - \mathfrak{c}_3) = a(z-\mathfrak{c}_3)^2 + b(z-\mathfrak{c}_3) + c$. We need to show that

$$(\mathrm{T}_{n,0}g)(z) = g\left(\frac{z-\mathfrak{c}_3}{\alpha_n} + \mathfrak{c}_3 \right) + \frac{a\gamma_n}{\alpha_{n-1}} \mathrm{U}_2\left(\frac{z-\mathfrak{c}_3}{\alpha_n} + \mathfrak{c}_3 \right), \tag{9.37}$$

$$(\mathrm{T}_{n,1}g)(z) = \frac{\gamma_n}{\alpha_n} \left(\frac{a(\alpha_n + \alpha\alpha_{n-1})}{\alpha_{n-1}^2} (z - \mathfrak{c}_3) + b \right), \tag{9.38}$$

$$(\mathrm{T}_{n,2}g)(z) = \frac{a\gamma_n\gamma_{n-1}}{\alpha_{n-1}^2}, \tag{9.39}$$

where $\mathrm{T}_{n,k}f$ is defined by (1.36)–(1.42). Note that

$$(\mathrm{T}_{n,0}g)(z) = \frac{\alpha a}{\alpha_n\alpha_{n-1}} (z-\mathfrak{c}_3)^2 + \frac{b}{\alpha_n}(z-\mathfrak{c}_3) + c + \frac{4a(1-\alpha^2)\gamma_n}{\alpha_{n-1}} \mathfrak{c}_1\mathfrak{c}_2. \tag{9.40}$$

We proceed by induction on n. Setting $n = 0$ in (9.37)–(9.39), we obtain $\mathrm{T}_{0,0}g = g$ and $\mathrm{T}_{0,1}g = 0 = \mathrm{T}_{0,2}g$. Next, suppose that (9.37)–(9.39) hold for all positive integers up to a fixed n. Then, by (1.42), we have

$$\mathrm{T}_{n+1,0}g = \mathrm{S}(\mathrm{T}_{n,0}g) - \frac{\gamma_{n+1}}{\alpha_{n+1}}\mathrm{U}_1\mathrm{D}(\mathrm{T}_{n,0}g), \tag{9.41}$$

$$\mathrm{T}_{n+1,1}g = \mathrm{S}(\mathrm{T}_{n,1}g) - \frac{\gamma_n}{\alpha_n}\mathrm{U}_1\mathrm{D}(\mathrm{T}_{n,1}g) + \frac{1}{\alpha_{n+1}}\mathrm{D}(\mathrm{T}_{n,0}g), \tag{9.42}$$

$$\mathrm{T}_{n+1,2}g = \mathrm{S}(\mathrm{T}_{n,2}g) + \frac{1}{\alpha_n}\mathrm{D}(\mathrm{T}_{n,1}g). \tag{9.43}$$

Using the identities

$$\mathrm{S}\left((z-\mathfrak{c}_3)^2\right) = (2\alpha^2-1)(z-\mathfrak{c}_3)^2 + 4(1-\alpha^2)\mathfrak{c}_1\mathfrak{c}_2,$$

$$\mathrm{D}\left((z-\mathfrak{c}_3)^2\right) = 2\alpha(z-\mathfrak{c}_3), \quad \mathrm{S}\left((z-\mathfrak{c}_3)\right) = \alpha(z-\mathfrak{c}_3),$$

we find

$$\mathrm{S}(\mathrm{T}_{n,0}g)(z) = \frac{\alpha(2\alpha^2-1)a}{\alpha_n\alpha_{n-1}}(z-\mathfrak{c}_3)^2 + \frac{\alpha b}{\alpha_n}(z-\mathfrak{c}_3) + c$$
$$+ \frac{4a(1-\alpha^2)(\alpha+\alpha_n\gamma_n)}{\alpha_n\alpha_{n-1}}\mathfrak{c}_1\mathfrak{c}_2$$

and

$$\mathrm{S}(\mathrm{T}_{n,1}g)(z) = \frac{\gamma_n}{\alpha_n}\left(\frac{\alpha a(\alpha_n+\alpha\alpha_{n-1})}{\alpha_{n1}^2}(z-\mathfrak{c}_3)+b\right),$$

$$\mathrm{D}(\mathrm{T}_{n,0}g)(z) = \frac{1}{\alpha_n}\left(\frac{2\alpha^2 a}{\alpha_{n-1}}(z-\mathfrak{c}_3)+b\right),$$

$$\mathrm{D}(\mathrm{T}_{n,1}g)(z) = \frac{a\gamma_n(\alpha_n+\alpha\alpha_{n-1})}{\alpha_n\alpha_{n-1}^2}.$$

Therefore, from (9.41) and using (1.11)–(1.18), we obtain

$$(\mathrm{T}_{n+1,0}g)(z) = \frac{\alpha a}{\alpha_n\alpha_{n-1}}\left(2\alpha^2-1+2\alpha\frac{(1-\alpha^2)\gamma_{n+1}}{\alpha_{n+1}}\right)(z-\mathfrak{c}_3)^2$$
$$+ \frac{b(\alpha\alpha_{n+1}+(1-\alpha^2)\gamma_{n+1})}{\alpha_n\alpha_{n+1}}(z-\mathfrak{c}_3)+c$$
$$+ \frac{4a(1-\alpha^2)(\alpha+\alpha_n\gamma_n)}{\alpha_n\alpha_{n-1}}\mathfrak{c}_1\mathfrak{c}_2$$
$$= \frac{\alpha a}{\alpha_n\alpha_{n+1}}(z-\mathfrak{c}_3)^2 + \frac{b}{\alpha_{n+1}}(z-\mathfrak{c}_3) + c + \frac{4a(1-\alpha^2)\gamma_{n+1}}{\alpha_n}\mathfrak{c}_1\mathfrak{c}_2.$$

Hence (9.37) holds for all n. Similarly, from (9.42), and using again (1.11)–(1.18) and the identity $\alpha_{n+1}\gamma_n(\alpha_n+\alpha\alpha_{n-1})+2\alpha^2\alpha_n=\alpha_{n-1}\gamma_{n+1}(\alpha_{n+1}+\alpha\alpha_n)$, we get

$$\begin{aligned}(\mathrm{T}_{n+1,1}g)(z) &= \frac{a}{\alpha_n\alpha_{n-1}}\left(\frac{\gamma_n(\alpha_n+\alpha\alpha_{n-1})}{\alpha_n}+\frac{2\alpha^2}{\alpha_{n+1}}\right)(z-\mathfrak{c}_3)\\ &\quad+\frac{b}{\alpha_n}\left(\gamma_n+\frac{1}{\alpha_{n+1}}\right)\\ &= \frac{\gamma_{n+1}}{\alpha_{n+1}}\left(\frac{a(\alpha_{n+1}+\alpha\alpha_n)}{\alpha_n^2}(z-\mathfrak{c}_3)+b\right).\end{aligned}$$

Hence (9.38) holds for $n\in\mathbb{N}_0$. Finally, from (9.43) it is obvious that (9.39) holds with n replaced by $n+1$ and, consequently, it holds for all n. Therefore, (9.37)–(9.39) hold and so (9.34) is proved. □

Theorem 9.1 (Functional Rodrigues Formula)**.** *Consider the lattice*

$$x(s):=\mathfrak{c}_1q^{-s}+\mathfrak{c}_2q^{s}+\mathfrak{c}_3.$$

Let $\mathbf{u}\in\mathcal{P}'$ *and suppose that there exists an* x*-admissible pair* (ϕ,ψ) *such that* $\mathbf{u}$ *fulfills the functional equation* (9.2). *Set*

$$d_n:=\tfrac{1}{2}\phi''\gamma_n+\psi'\alpha_n,\quad e_n:=\phi'(\mathfrak{c}_3)\gamma_n+\psi(\mathfrak{c}_3)\alpha_n,\quad n\in\mathbb{N}_0. \tag{9.44}$$

Then

$$R_n\mathbf{u}=\mathbf{D}^n\mathbf{u}^{[n]},\quad n\in\mathbb{N}_0, \tag{9.45}$$

where $\mathbf{u}^{[n]}$ *is the functional on* $\mathcal{P}'$ *defined by* (9.7) *and* $\{R_n\}_{n\geq0}$ *is a simple set of polynomials given by*

$$R_{n+1}(z)=(a_nz-s_n)R_n(z)-t_nR_{n-1}(z), \tag{9.46}$$

with initial conditions $R_{-1}=0$ *and* $R_0=1$, *and*

$$a_n:=-\frac{\alpha\, d_{2n}d_{2n-1}}{d_{n-1}}, \tag{9.47}$$

$$s_n:=a_n\left(\mathfrak{c}_3+\frac{\gamma_ne_{n-1}}{d_{2n-2}}-\frac{\gamma_{n+1}e_n}{d_{2n}}\right), \tag{9.48}$$

$$t_n:=a_n\frac{\alpha\,\gamma_nd_{2n-2}}{d_{2n-1}}\phi^{[n-1]}\left(\mathfrak{c}_3-\frac{e_{n-1}}{d_{2n-2}}\right), \tag{9.49}$$

$\phi^{[n-1]}$ *being given by* (9.15). (*It is understood that* $a_0:=-\alpha d$ *and* $s_0:=\alpha e$.)

Proof. The proof is by induction on n. For $n = 0$, (9.45) is trivial. For $n = 1$, (9.45) follows from (9.28), since $R_1 = -\alpha\psi$. Assume now (induction hypothesis) that (9.45) holds for two consecutive nonnegative integer numbers, i.e., the relations

$$R_{n-1}\mathbf{u} = \mathbf{D}^{n-1}\mathbf{u}^{[n-1]}, \quad R_n\mathbf{u} = \mathbf{D}^n\mathbf{u}^{[n]} \tag{9.50}$$

hold for some fixed $n \in \mathbb{N}$. We need to prove that $R_{n+1}\mathbf{u} = \mathbf{D}^{n+1}\mathbf{u}^{[n+1]}$. Notice first that, by (9.14) and (9.44), we have

$$\psi^{[k]}(z) = d_{2k}(z - \mathfrak{c}_3) + e_k, \quad k \in \mathbb{N}_0. \tag{9.51}$$

By (9.28) and the Leibniz formula, we may write

$$\begin{aligned}\mathbf{D}^{n+1}\mathbf{u}^{[n+1]} &= \mathbf{D}^n\mathbf{D}\mathbf{u}^{[n+1]} = -\alpha\mathbf{D}^n(\psi^{[n]}\mathbf{u}^{[n]}) \\ &= -\alpha\mathrm{T}_{n,0}\psi^{[n]}\mathbf{D}^n\mathbf{u}^{[n]} - \alpha\mathrm{T}_{n,1}\psi^{[n]}\mathbf{D}^{n-1}\mathbf{S}\mathbf{u}^{[n]}.\end{aligned}$$

From (9.36) we have $\mathrm{T}_{n,1}\psi^{[n]} = d_{2n}\gamma_n/\alpha_n$, and so, using also (9.50),

$$\mathbf{D}^{n-1}\mathbf{S}\mathbf{u}^{[n]} = -\frac{\alpha_n}{\alpha d_{2n}\gamma_n}\left(\mathbf{D}^{n+1}\mathbf{u}^{[n+1]} + \alpha\left(\mathrm{T}_{n,0}\psi^{[n]}\right)R_n\mathbf{u}\right). \tag{9.52}$$

Shifting n into $n-1$, and using again the induction hypothesis (9.50), we obtain

$$\mathbf{D}^{n-2}\mathbf{S}\mathbf{u}^{[n-1]} = -\frac{\alpha_{n-1}}{\alpha d_{2n-2}\gamma_{n-1}}\left(R_n + \alpha\left(\mathrm{T}_{n-1,0}\psi^{[n-1]}\right)R_{n-1}\right)\mathbf{u}. \tag{9.53}$$

Next, using (9.28), (1.26), and (9.29), we deduce

$$\begin{aligned}\mathbf{D}^{n+1}\mathbf{u}^{[n+1]} &= -\alpha\mathbf{D}^n\left(\psi^{[n]}\mathbf{u}^{[n]}\right) = -\alpha\mathbf{D}^{n-1}\left(\mathbf{D}(\psi^{[n]}\mathbf{u}^{[n]})\right) \\ &= -\mathbf{D}^{n-1}\left(\left(\alpha\mathrm{S}\psi^{[n]} - \mathrm{U}_1\mathrm{D}\psi^{[n]}\right)\mathbf{D}\mathbf{u}^{[n]} + \mathrm{D}\psi^{[n]}\mathbf{S}\mathbf{u}^{[n]}\right) \\ &= \mathbf{D}^{n-1}\left(\xi_2(\cdot;n)\mathbf{u}^{[n-1]}\right),\end{aligned} \tag{9.54}$$

where $\xi_2(\cdot;n)$ is a polynomial of degree 2, given by

$$\xi_2(z;n) = \alpha^2\left(\psi^{[n-1]}\mathrm{S}\psi^{[n]} + \phi^{[n-1]}\mathrm{D}\psi^{[n]}\right)(z). \tag{9.55}$$

The following identities may be proved by a straightforward computation:

$$d_{2n-1} - \alpha d_{2n-2} = a^{[n-1]},$$

$$d_{2n-2}\left(e_n - 2\alpha\mathfrak{c}_3 d_{2n}\right) + d_{2n}\left(b^{[n-1]} + \alpha e_{n-1}\right) = 2d_{2n-1}(\alpha e_n - \mathfrak{c}_3 d_{2n}).$$

(The second one is achieved by using equation (9.21).) Using these relations, together with (9.16), (9.51), (9.8), and (9.9), we deduce

$$\xi_2(z;n) = \alpha^2 d_{2n}d_{2n-1}z^2 + 2\alpha^2 d_{2n-1}(\alpha e_n - \mathfrak{c}_3 d_{2n})z \tag{9.56}$$
$$+ \alpha^2\left(d_{2n}c^{[n-1]} + (e_{n-1} - \mathfrak{c}_3 d_{2n-2})(e_n - \alpha\mathfrak{c}_3 d_{2n})\right).$$

Since $\deg \xi_2(\cdot;n) = 2$, using again the Leibniz formula, we may write

$$\mathbf{D}^{n-1}\left(\xi_2(\cdot;n)\mathbf{u}^{[n-1]}\right) = \mathrm{T}_{n-1,0}\xi_2(\cdot;n)\mathbf{D}^{n-1}\mathbf{u}^{[n-1]} \tag{9.57}$$
$$+ \mathrm{T}_{n-1,1}\xi_2(\cdot;n)\mathbf{D}^{n-2}\mathbf{S}\mathbf{u}^{[n-1]} \tag{9.58}$$
$$+ \mathrm{T}_{n-1,2}\xi_2(\cdot;n)\mathbf{D}^{n-3}\mathbf{S}^2\mathbf{u}^{[n-1]}.$$

Since, by (9.36), $\mathrm{T}_{n-1,2}\xi_2(\cdot;n) = \alpha^2\gamma_{n-1}\gamma_{n-2}d_{2n}d_{2n-1}/\alpha_{n-2}^2$, combining equations (9.57), (9.54), (9.53), and (9.50), we obtain

$$\mathbf{D}^{n-3}\mathbf{S}^2\mathbf{u}^{[n-1]} = \frac{\alpha_{n-2}^2}{\alpha^2\gamma_{n-1}\gamma_{n-2}d_{2n}d_{2n-1}}\Big(\mathbf{D}^{n+1}\mathbf{u}^{[n+1]} - \left(\mathrm{T}_{n-1,0}\xi_2(\cdot;n)\right)R_{n-1}\mathbf{u} \tag{9.59}$$
$$+ \frac{\alpha_{n-1}\mathrm{T}_{n-1,1}\xi_2(\cdot;n)}{\alpha\gamma_{n-1}d_{2n-2}}\left(R_n + \alpha\left(\mathrm{T}_{n-1,0}\psi^{[n-1]}\right)R_{n-1}\right)\mathbf{u}\Big).$$

On the other hand, by (9.29),

$$\mathbf{S}\mathbf{u}^{[n]} = \eta_2(\cdot;n)\mathbf{u}^{[n-1]}, \quad \eta_2(z;n) := -\alpha\left(\alpha\phi^{[n-1]} + \mathfrak{U}_1\psi^{[n-1]}\right)(z). \tag{9.60}$$

Taking into account that $\eta_2(\cdot;n)$ is a polynomial of degree at most two, by the Leibniz formula and (9.50), we may write

$$\mathbf{D}^{n-1}\mathbf{S}\mathbf{u}^{[n]} = \mathbf{D}^{n-1}\left(\eta_2(\cdot;n)\mathbf{u}^{[n-1]}\right) \tag{9.61}$$
$$= \left(\mathrm{T}_{n-1,0}\eta_2(\cdot;n)\right)R_{n-1}\mathbf{u} + \mathrm{T}_{n-1,1}\eta_2(\cdot;n)\mathbf{D}^{n-2}\mathbf{S}\mathbf{u}^{[n-1]}$$
$$+ \mathrm{T}_{n-1,2}\eta_2(\cdot;n)\mathbf{D}^{n-3}\mathbf{S}^2\mathbf{u}^{[n-1]}.$$

Note that $\eta_2(\cdot;n)$ is given explicitly by

$$\eta_2(z;n) = \alpha(\alpha d_{2n-1} - d_{2n})z^2 - \alpha\left(\alpha b^{[n-1]} + (\alpha^2-1)(e_{n-1} - 2\mathfrak{c}_3 d_{2n-2})\right)z \tag{9.62}$$
$$- \alpha\left(\alpha c^{[n-1]} + \beta(\alpha+1)(e_{n-1} - \mathfrak{c}_3 d_{2n-2})\right).$$

Hence, using (9.36), we get

$$\mathrm{T}_{n-1,2}\eta_2(\cdot;n) = \alpha\gamma_{n-1}\gamma_{n-2}(\alpha d_{2n-1} - d_{2n})/\alpha_{n-2}^2.$$

Therefore, substituting (9.52), (9.53), and (9.59) in (9.61), we obtain

$$\mathbf{D}^{n+1}\mathbf{u}^{[n+1]} = \left(A(\cdot;n)R_n + B(\cdot;n)R_{n-1}\right)\mathbf{u}, \tag{9.63}$$

where $A(\cdot;n)$ and $B(\cdot;n)$ are polynomials depending on n, given by

$$\begin{aligned}\epsilon_n A(z;n) &= \frac{\alpha_n\left(\mathrm{T}_{n,0}\psi^{[n]}\right)(z)}{\gamma_n d_{2n}} - \frac{\alpha_{n-1}\left(\mathrm{T}_{n-1,1}\eta_2\right)(z;n)}{\alpha\gamma_{n-1}d_{2n-2}} \\ &\quad + \frac{\alpha_{n-1}(\alpha d_{2n-1}-d_{2n})\left(\mathrm{T}_{n-1,1}\xi_2\right)(z;n)}{\alpha^2\gamma_{n-1}d_{2n}d_{2n-1}d_{2n-2}}\end{aligned} \tag{9.64}$$

and

$$\begin{aligned}\epsilon_n B(z;n) &= \left(\mathrm{T}_{n-1,0}\eta_2\right)(z;n) - \frac{\alpha_{n-1}\left(\mathrm{T}_{n-1,0}\psi^{[n-1]}\right)(z)\left(\mathrm{T}_{n-1,1}\eta_2\right)(z;n)}{\gamma_{n-1}d_{2n-2}} \\ &\quad + \frac{(d_{2n}-\alpha d_{2n-1})\left(\mathrm{T}_{n-1,0}\xi_2\right)(z;n)}{\alpha d_{2n}d_{2n-1}} \\ &\quad + \frac{\alpha_{n-1}(\alpha d_{2n-1}-d_{2n})\left(\mathrm{T}_{n-1,1}\xi_2\right)(z;n)\left(\mathrm{T}_{n-1,0}\psi^{[n-1]}\right)(z)}{\alpha\gamma_{n-1}d_{2n}d_{2n-1}d_{2n-2}},\end{aligned} \tag{9.65}$$

where

$$\epsilon_n := \frac{d_{2n}-\alpha d_{2n-1}}{\alpha d_{2n}d_{2n-1}} - \frac{\alpha_n}{\alpha\gamma_n d_{2n}} = -\frac{d_{n-1}}{\alpha\gamma_n d_{2n}d_{2n-1}}.$$

Note that

$$\gamma_n d_{2n} - (\alpha+\alpha_n)d_{2n-1} = -d_{n-1}.$$

We claim that

$$\begin{aligned}A(z;n) &= -\alpha\frac{d_{2n}d_{2n-1}}{d_{n-1}}(z-\mathfrak{c}_3) + \frac{\alpha\gamma_n d_{2n}d_{2n-1}e_{n-1}}{d_{2n-2}d_{n-1}} - \frac{\alpha\gamma_{n+1}d_{2n-1}e_n}{d_{n-1}} \\ &= a_n z - s_n,\end{aligned} \tag{9.66}$$

$$B(z;n) = \alpha^2\frac{\gamma_n d_{2n}d_{2n-2}}{d_{n-1}}\phi^{[n-1]}\left(\mathfrak{c}_3 - \frac{e_{n-1}}{d_{2n-2}}\right) = -t_n, \tag{9.67}$$

where a_n, s_n, and t_n are given by (9.47)–(9.49). Indeed, by (9.22) and (9.23),

$$a^{[n-1]} = d_{2n-1} - \alpha d_{2n-2}, \quad n\in\mathbb{N}, \tag{9.68}$$

$$b^{[n-1]} = e_n - \alpha e_{n-1} - 2\mathfrak{c}_3 a^{[n-1]}, \quad n\in\mathbb{N}, \tag{9.69}$$

$$d_{2n} - 2\alpha d_{2n-1} + d_{2n-2} = 0, \quad n\in\mathbb{N}. \tag{9.70}$$

By (9.56), (9.62), and (9.51), and applying (9.34)–(9.36) together with (9.69), we obtain

$$(\mathrm{T}_{n,0}\psi^{[n]})(z) = \frac{d_{2n}}{\alpha_n}(z - \mathfrak{c}_3) + e_n, \tag{9.71}$$

$$(\mathrm{T}_{n-1,1}\xi_2)(z;n) = \frac{\gamma_{n-1}}{\alpha_{n-1}}\left(\frac{\alpha^2 d_{2n}d_{2n-1}(\alpha_{n-1}+\alpha\alpha_{n-2})}{\alpha_{n-2}^2}(z-\mathfrak{c}_3) + 2\alpha^3 e_n d_{2n-1}\right), \tag{9.72}$$

and

$$(\mathrm{T}_{n-1,1}\eta_2)(z;n) = \frac{\gamma_{n-1}}{\alpha_{n-1}}\left(\frac{\alpha(\alpha d_{2n-1}-d_{2n})(\alpha_{n-1}+\alpha\alpha_{n-2})}{\alpha_{n-2}^2}(z-\mathfrak{c}_3) + \alpha(e_{n-1}-\alpha e_n)\right). \tag{9.73}$$

Similarly,

$$\begin{aligned}(\mathrm{T}_{n-1,1}\eta_2)(z;n) =& \frac{\alpha^2(\alpha d_{2n-1}-d_{2n})}{\alpha_{n-1}\alpha_{n-2}}(z-\mathfrak{c}_3)^2 + \frac{\alpha(e_{n-1}-\alpha e_n)}{\alpha_{n-1}}(z-\mathfrak{c}_3)\\ &+ \eta_2(\mathfrak{c}_3;n) + \frac{4\alpha(1-\alpha^2)\gamma_{n-1}(\alpha d_{2n-1}-d_{2n})}{\alpha_{n-2}}\mathfrak{c}_1\mathfrak{c}_2, \end{aligned} \tag{9.74}$$

$$\begin{aligned}(\mathrm{T}_{n-1,1}\xi_2)(z;n) =& \frac{\alpha^3 d_{2n-1}d_{2n}}{\alpha_{n-1}\alpha_{n-2}}(z-\mathfrak{c}_3)^2 + \frac{2\alpha^3 e_n d_{2n-1}}{\alpha_{n-1}}(z-\mathfrak{c}_3) + \xi_2(\mathfrak{c}_3;n)\\ &+ \frac{4\alpha^2(1-\alpha^2)\gamma_{n-1}d_{2n-1}d_{2n}}{\alpha_{n-2}}\mathfrak{c}_1\mathfrak{c}_2. \end{aligned} \tag{9.75}$$

Using (9.71), (9.72), and (9.73) together with the identity $\alpha_n + \alpha\gamma_n = \gamma_{n+1}$ (this last one follows from (1.11) and (1.12)), we deduce from (9.64) that

$$\begin{aligned}\epsilon_n A(z;n) &= \frac{1}{\gamma_n}(z-\mathfrak{c}_3) - \frac{e_{n-1}}{d_{2n-2}} + \frac{\alpha_n e_n}{\gamma_n d_{2n}} + \alpha e_n \frac{2\alpha d_{2n-1}-d_{2n}}{d_{2n}d_{2n-2}}\\ &= \frac{1}{\gamma_n}(z-\mathfrak{c}_3) - \frac{e_{n-1}}{d_{2n-2}} + \frac{\gamma_{n+1}e_n}{\gamma_n d_{2n}},\end{aligned}$$

and (9.66) follows. From (9.71)–(9.75) it is straightforward to verify that (9.65) reduces to

$$\begin{aligned}\epsilon_n B(z;n) = \eta_2(\mathfrak{c}_3;n) &+ \frac{(d_{2n}-\alpha d_{2n-1})\xi_2(\mathfrak{c}_3;n)}{\alpha d_{2n}d_{2n-1}} + \frac{\alpha(\alpha e_n - e_{n-1})e_{n-1}}{d_{2n-2}}\\ &+ \frac{2\alpha^2(\alpha d_{2n-1}-d_{2n})e_n e_{n-1}}{d_{2n}d_{2n-2}}.\end{aligned}$$

Moreover, from the definitions of $\xi(.;n)$ and $\eta_2(.;n)$ given in (9.55) and (9.60), we deduce

$$\xi(\mathfrak{c}_3;n) = \alpha^2(e_n e_{n-1} + \phi^{[n-1]}(\mathfrak{c}_3)d_{2n}), \quad \eta_2(\mathfrak{c}_3;n) = -\alpha^2\phi^{[n-1]}(\mathfrak{c}_3).$$

Consequently, by (9.68), we show that

$$\epsilon_n B(z;n) = -\alpha\frac{d_{2n-2}}{d_{2n-1}}\phi^{[n-1]}(\mathfrak{c}_3) + \alpha e_{n-1}\left(\frac{e_n}{d_{2n-1}} - \frac{e_{n-1}}{d_{2n-2}}\right).$$

Therefore, using successively (9.68) and (9.69), we obtain

$$\begin{aligned}
B(z;n) &= \frac{\alpha^2\gamma_n d_{2n}d_{2n-2}}{d_{n-1}}\left(d_{2n-1}\frac{e_{n-1}^2}{d_{2n-2}^2} - e_n\frac{e_{n-1}}{d_{2n-2}} + \phi^{[n-1]}(\mathfrak{c}_3)\right) \\
&= \frac{\alpha^2\gamma_n d_{2n}d_{2n-2}}{d_{n-1}}\left((a^{[n-1]} + \alpha d_{2n-2})\frac{e_{n-1}^2}{d_{2n-2}^2}\right. \\
&\quad \left. -(b^{[n-1]} + \alpha e_{n-1} + 2\mathfrak{c}_3 a^{[n-1]})\frac{e_{n-1}}{d_{2n-2}} + \phi^{[n-1]}(\mathfrak{c}_3)\right) \\
&= \frac{\alpha^2\gamma_n d_{2n}d_{2n-2}}{d_{n-1}}\phi^{[n-1]}\left(\mathfrak{c}_3 - \frac{e_{n-1}}{d_{2n-2}}\right),
\end{aligned}$$

and (9.67) is proved. It follows from (9.66) and (9.67) that (9.63) reduces to $\mathbf{D}^{n+1}\mathbf{u}^{[n+1]} = R_{n+1}\mathbf{u}$, which completes the proof. □

9.1.3 The Regularity of $\mathbf{u}^{[k]}$

The next result follows easily, see Lemma 7.2.

Lemma 9.3. *Suppose that $\mathbf{u} \in \mathcal{P}'$ is regular and satisfies the functional differential equation (9.2), where $\phi \in \mathcal{P}_2$ and $\psi \in \mathcal{P}_1$. Assume that at least one of the polynomials ϕ and ψ is nonzero. Then neither ϕ nor ψ is the zero polynomial and*

$$\deg\psi = 1.$$

Lemma 9.4 below gives the regularity of $\mathbf{u}^{[k]}$.

Lemma 9.4. *Suppose that $\mathbf{u} \in \mathcal{P}'$ is regular and satisfies (9.2), where $(\phi,\psi) \in \mathcal{P}_2 \times \mathcal{P}_1 \setminus \{(0,0)\}$. Then (ϕ,ψ) is an x-admissible pair and $\mathbf{u}^{[k]}$, $k \in \mathbb{N}$, is regular. Moreover, if $\{P_n\}_{n\geq 0}$ is the monic OPS with respect to $\mathbf{u}$, then $\left\{P_n^{[k]}\right\}_{n\geq 0}$ is the monic OPS with respect to $\mathbf{u}^{[k]}$.*

Proof. Set $\phi(z) = az^2 + bz + c$ and $\psi(z) = dz + e$. By Lemma 9.3, both ϕ and ψ are nonzero polynomials, and $d \neq 0$. If $\deg \phi \in \{0, 1\}$ then $d_n = d\alpha_n \neq 0$ for each $n \in \mathbb{N}$, and so the pair (ϕ, ψ) is x-admissible. Assume now that $\deg \phi = 2$. Then $d_n = a\gamma_n + d\alpha_n$, with $a \neq 0$. To prove that $d_n \neq 0$, we start by showing that

$$\left\langle \mathbf{u}, \left(\mathrm{U}_2\psi\mathrm{D}P_n^{[1]} + \phi\mathrm{S}P_n^{[1]}\right) P_{n+2}\right\rangle$$
$$= -\left\langle \mathbf{u}^{[1]}, \left(\mathrm{S}P_{n+2} + \alpha^{-1}\mathrm{U}_1\mathrm{D}P_{n+2}\right) P_n^{[1]}\right\rangle. \tag{9.76}$$

Indeed, we have

$$\left\langle \mathbf{u}, \left(\mathrm{U}_2\psi\mathrm{D}P_n^{[1]} + \phi\mathrm{S}P_n^{[1]}\right) P_{n+2}\right\rangle = \left\langle \mathrm{U}_2\psi\mathbf{u}, P_{n+2}\mathrm{D}P_n^{[1]}\right\rangle + \left\langle \phi\mathbf{u}, P_{n+2}\mathrm{S}P_n^{[1]}\right\rangle$$
$$= \left\langle \mathrm{U}_2\psi\mathbf{u}, \mathrm{D}\left((\mathrm{S}P_{n+2} - \alpha^{-1}\mathrm{U}_1\mathrm{D}P_{n+2})P_n^{[1]}\right) - \alpha^{-1}\mathrm{S}\left(P_n^{[1]}\mathrm{D}P_{n+2}\right)\right\rangle$$
$$+ \left\langle \phi\mathbf{u}, \mathrm{S}\left((\mathrm{S}P_{n+2} - \alpha^{-1}\mathrm{U}_1\mathrm{D}P_{n+2})P_n^{[1]}\right) - \alpha^{-1}\mathrm{U}_2\mathrm{D}\left(P_n^{[1]}\mathrm{D}P_{n+2}\right)\right\rangle$$
$$= -\left\langle \mathbf{u}^{[1]}, \left(\mathrm{S}P_{n+2} - \alpha^{-1}\mathrm{U}_1\mathrm{D}P_{n+2}\right) P_n^{[1]}\right\rangle$$
$$- \alpha^{-1}\left\langle \mathbf{S}(\mathrm{U}_2\psi\mathbf{u}) - \mathbf{D}(\mathrm{U}_2\phi\mathbf{u}), P_n^{[1]}\mathrm{D}P_{n+2}\right\rangle,$$

where the second equality holds by (1.45) and (1.44). Therefore, using (9.30) for $n = 0$, we obtain (9.76). Now, on the one hand, $\mathrm{U}_2\psi\mathrm{D}P_n^{[1]} + \phi\mathrm{S}P_n^{[1]}$ is a polynomial of degree at most $n + 2$, being the coefficient of z^{n+2} equal to $(\alpha^2 - 1)d\gamma_n + a\alpha_n$. Hence, since the relations

$$(\alpha^2 - 1)d\gamma_n + a\alpha_n = d_{n+1} - \alpha d_n = \alpha d_n - d_{n-1}, \quad n \in \mathbb{N},$$

hold, we get

$$\mathrm{U}_2\psi\mathrm{D}P_n^{[1]} + \phi\mathrm{S}P_n^{[1]} = (\alpha d_n - d_{n-1})\, z^{n+2} + (\text{lower degree terms}), \quad n \in \mathbb{N}.$$

Consequently,

$$\left\langle \mathbf{u}, \left(\mathrm{U}_2\psi\mathrm{D}P_n^{[1]} + \phi\mathrm{S}P_n^{[1]}\right) P_{n+2}\right\rangle = (\alpha d_n - d_{n-1})\langle \mathbf{u}, P_{n+2}^2\rangle. \tag{9.77}$$

On the other hand, since

$$\mathrm{S}P_{n+2} + \alpha^{-1}\mathrm{U}_1\mathrm{D}P_{n+2} = \sum_{j=0}^{n+2} c_{n,j}P_j^{[1]}, \quad c_{n,j} \in \mathbb{C},$$

and using

$$\left\langle \mathbf{u}^{[k]}, P_n^{[k]}P_m^{[k]}\right\rangle = \alpha\frac{d_n^{[k-1]}}{\gamma_{n+1}}\langle \mathbf{u}, (P_{n+1}^{[k-1]})^2\rangle\delta_{n,m}, \quad 0 \leq m \leq n, \quad n \in \mathbb{N}_0, \tag{9.78}$$

$k=1$, we obtain

$$\left\langle \mathbf{u}^{[1]}, \left(\mathrm{S}P_{n+2}+\alpha^{-1}\mathrm{U}_1\mathrm{D}P_{n+2}\right)P_n^{[1]}\right\rangle = \frac{\alpha c_{n,n}d_n}{\gamma_{n+1}}\langle \mathbf{u}, P_{n+1}^2\rangle, \quad n\in\mathbb{N}. \tag{9.79}$$

Substituting (9.77) and (9.79) into (9.76), and since $C_{n+2}=\langle \mathbf{u}, P_{n+2}^2\rangle/\langle \mathbf{u}, P_{n+1}^2\rangle$, we deduce

$$\alpha\left(1+\frac{c_{n,n}}{\gamma_{n+1}C_{n+2}}\right)d_n = d_{n-1}, \quad n\in\mathbb{N}.$$

Therefore, since $d_0=d\neq 0$, we conclude that $d_n\neq 0$, $n\in\mathbb{N}_0$, and so (ϕ,ψ) is an x-admissible pair. Consequently, by (9.78) for $k=1$, $\{P_n^{[1]}\}_{n\geq 0}$ is the monic OPS with respect to $\mathbf{u}^{[1]}$. This proves the last statement in the lemma for $k=1$. Since

$$d_n^{[k]} = a^{[k]}\gamma_n + \alpha_n d^{[k]}, \quad k,n\in\mathbb{N}_0,$$

it is easy to see, using (9.22)–(9.25), that $d_n^{[k]}=d_{n+2k}$ for all $n,k\in\mathbb{N}_0$. Thus the last sentence of the lemma follows from (9.78). □

9.2 The Regularity Conditions

In this section we state our main results, giving the analog of Theorem 7.2 on lattices. Thus, once a lattice as in (1.9) is fixed, we state necessary and sufficient conditions so that a functional $\mathbf{u}\in\mathcal{P}'$ satisfying (9.2) is regular. Furthermore, the monic OPS with respect to $\mathbf{u}$ is described.

9.2.1 The Lattice $x(s)=\mathfrak{c}_1 q^{-s}+\mathfrak{c}_2 q^s+\mathfrak{c}_3$

We start by considering a lattice $x(s)$ with $q\neq 1$, so that $x(s)$ is a q-quadratic or a q-linear lattice.

Theorem 9.2. *Consider the lattice*

$$x(s)=\mathfrak{c}_1 q^{-s}+\mathfrak{c}_2 q^s+\mathfrak{c}_3.$$

Suppose that $\mathbf{u}\in\mathcal{P}'$ *satisfies the functional differential* (9.2), *with* ϕ *and* ψ *given by* (7.2), *where at least one of these polynomials is nonzero. Let* d_n *and* e_n *be defined by* (9.44), *and* $\phi^{[n]}$ *and* $\psi^{[n]}$ *be given by* (9.14)–(9.15). *Then,* $\mathbf{u}$ *is regular if and only if*

$$d_n\neq 0, \quad \phi^{[n]}\left(\mathfrak{c}_3-\frac{e_n}{d_{2n}}\right)\neq 0, \quad n\in\mathbb{N}_0. \tag{9.80}$$

Moreover, under such conditions, the monic OPS $\{P_n\}_{n\geq 0}$ *with respect to* $\mathbf{u}$ *satisfies*

$$P_{n+1}(z) = (z - B_n)P_n(z) - C_nP_{n-1}(z), \tag{9.81}$$

with $P_{-1} = 0$ *and* $P_0 = 1$, *where the recurrence coefficients are given by*

$$B_n = \mathfrak{c}_3 + \frac{\gamma_n e_{n-1}}{d_{2n-2}} - \frac{\gamma_{n+1}e_n}{d_{2n}}, \tag{9.82}$$

$$C_{n+1} = -\frac{\gamma_{n+1}d_{n-1}}{d_{2n-1}d_{2n+1}}\phi^{[n]}\left(\mathfrak{c}_3 - \frac{e_n}{d_{2n}}\right). \tag{9.83}$$

In addition, the following functional Rodrigues formula holds:

$$P_n\mathbf{u} = k_n\mathbf{D}^n\mathbf{u}^{[n]}, \quad k_n := (-\alpha)^{-n}\prod_{j=1}^{n} d_{n+j-2}^{-1}. \tag{9.84}$$

Proof. Suppose that $\mathbf{u}$ is regular. By Lemma 9.4, (ϕ, ψ) is an x-admissible pair, meaning that the first condition in (9.80) holds, and $\left\{P_j^{[n]}\right\}_{j\geq 0}$ is the monic OPS with respect to $\mathbf{u}^{[n]}$, for each fixed n. Write the recurrence relation for $\left\{P_j^{[n]}\right\}_{j\geq 0}$ as

$$P_{j+1}^{[n]}(z) = (z - B_j^{[n]})P_j^{[n]}(z) - C_j^{[n]}P_{j-1}^{[n]}(z) \tag{9.85}$$

$(P_{-1}^{[n]} := 0)$, where $B_j^{[n]} \in \mathbb{C}$ and $C_{j+1}^{[n]} \in \mathbb{C}\setminus\{0\}$. Let us compute $C_1^{[n]}$. We first show that (for $n = 0$) the coefficient $C_1 \equiv C_1^{[0]}$, appearing in the recurrence relation for $\{P_j\}_{j\geq 0}$, is given by

$$C_1 = -\frac{1}{d\alpha + a}\phi\left(-\frac{e}{d}\right) = -\frac{1}{d_1}\phi\left(\mathfrak{c}_3 - \frac{e_0}{d_0}\right). \tag{9.86}$$

Indeed, taking $k = 0$ and $k = 1$ in the relation

$$\langle \mathbf{D}_x(\phi\mathbf{u}), z^k\rangle = \langle \mathbf{S}_x(\psi\mathbf{u}), z^k\rangle,$$

we obtain $0 = du_1 + eu_0$ and $au_2 + bu_1 + cu_0 = -d\alpha u_2 - (e\alpha + d\beta)u_1 - e\beta u_0$, where $u_k := \langle \mathbf{u}, z^k\rangle$, $k \in \mathbb{N}_0$. Therefore,

$$u_1 = -\frac{e}{d}u_0, \quad u_2 = -\frac{1}{d\alpha + a}\left(-(b + e\alpha)\frac{e}{d} + c\right)u_0. \tag{9.87}$$

On the other hand, since $P_1(z) = z - B_0^{[0]} = z - u_1/u_0$, we also have

$$C_1 = \frac{\langle \mathbf{u}, P_1^2 \rangle}{u_0} = \frac{u_2 u_0 - u_1^2}{u_0^2}$$

$$= \frac{u_2}{u_0} - \left(\frac{u_1}{u_0}\right)^2. \tag{9.88}$$

Substituting u_1 and u_2 given by (9.87) into (9.88) yields (9.86). Since the equation (9.27) fulfilled by $\mathbf{u}^{[n]}$ is of the same type as (9.2) fulfilled by $\mathbf{u}$, with the polynomials $\phi^{[n]}$ and $\psi^{[n]}$ in (9.27) playing the roles of ϕ and ψ in (9.2), $C_1^{[n]}$ may be obtained by replacing in (9.86) ϕ and $\psi(z) = dz + e$ by $\phi^{[n]}$ and $\psi^{[n]}(z) = d_{2n}(z - \mathfrak{c}_3) + e_n$, respectively. Hence,

$$C_1^{[n]} = -\frac{1}{d_{2n}\alpha + a^{[n]}} \phi^{[n]}\left(\mathfrak{c}_3 - \frac{e_n}{d_{2n}}\right)$$

$$= -\frac{1}{d_{2n+1}} \phi^{[n]}\left(\mathfrak{c}_3 - \frac{e_n}{d_{2n}}\right). \tag{9.89}$$

Since $\mathbf{u}^{[n]}$ is regular, then $C_1^{[n]} \neq 0$, and so the second condition in (9.80) holds.

Conversely, suppose that conditions (9.80) hold. Define a sequence of polynomials, $\{P_n\}_{n\geq 0}$, by the recurrence relation (9.81)–(9.83), with $P_{-1} := 0$. According to the hypothesis (9.80), $C_{n+1} \neq 0$. Therefore, the Favard theorem ensures that $\{P_n\}_{n\geq 0}$ is a monic OPS. To prove that $\mathbf{u}$ is regular we will show that $\{P_n\}_{n\geq 0}$ is the monic OPS with respect to $\mathbf{u}$. For that we only need to prove that

$$u_0 \neq 0, \quad \langle \mathbf{u}, P_n \rangle = 0, \quad n \in \mathbb{N}_0.$$

We start by showing that $u_0 \neq 0$. Indeed, suppose that $u_0 = 0$. Since (9.2) holds, then

$$\langle \mathbf{D}(\phi \mathbf{u}) - \mathbf{S}(\psi \mathbf{u}), z^n \rangle = 0.$$

This implies

$$d_n u_{n+1} + \sum_{j=0}^{n} a_{n,j} u_j = 0, \quad a_{n,j} \in \mathbb{C}. \tag{9.90}$$

Since $u_0 = 0$ and $d_n \neq 0$, from (9.90) we deduce $u_n = 0$, $n \in \mathbb{N}_0$, and so $\mathbf{u} = \mathbf{0}$, contrary to the hypothesis. Therefore $u_0 \neq 0$. Next, note that $P_n(z) = k_n R_n(z)$, where R_n is defined by (9.46) and

$$k_n^{-1} := (-\alpha)^n \prod_{j=1}^{n} d_{n+j-2}.$$

By the Rodrigues formula (9.45) given in Theorem 9.1, we obtain

$$\begin{aligned}\langle \mathbf{u}, P_n\rangle &= k_n\langle \mathbf{u}, R_n\rangle = k_n\langle R_n\mathbf{u}, 1\rangle \\ &= k_n\left\langle \mathrm{D}^n\mathbf{u}^{[n]}, 1\right\rangle = (-1)^n k_n\left\langle \mathbf{u}^{[n]}, \mathrm{D}^n 1\right\rangle = 0, n\in\mathbb{N}.\end{aligned}$$

Thus $\mathbf{u}$ is regular. The remaining statements of the theorem follow from the Theorem 9.1. □

Remark 9.1. *It is important to highlight that the Rodrigues formula holds for a functional* $\mathbf{u}$ *satisfying a functional equation, even if* $\mathbf{u}$ *is not regular, provided that the pair* (ϕ, ψ) *appearing in that* x*-functional equation forms an* x*-admissible pair. This fact is a consequence of Theorem 9.1.*

9.2.2 The Lattice $x(s) = \mathfrak{c}_4 s^2 + \mathfrak{c}_5 s + \mathfrak{c}_6$

We consider now a lattice $x(s)$ for $q = 1$, so that $x(s)$ is a quadratic or a linear lattice. Recall that, here, $\mathfrak{c}_4 = 4\beta$. We just give the results for this situation, since the techniques and computations are very similar to the ones presented for the case $q \neq 1$. For the lattice $x(s) = \mathfrak{c}_4 s^2 + \mathfrak{c}_5 s + \mathfrak{c}_6$ the system of difference equations (9.17)–(9.21) become

$$\begin{aligned}
a^{[n+1]} &= a^{[n]},\\
d^{[n+1]} &= 2a^{[n]} + d^{[n]},\\
b^{[n+1]} &= b^{[n]} + 6\beta(a^{[n]} + d^{[n]}),\\
e^{[n+1]} &= e^{[n]} + b^{[n]} + \beta(2a^{[n]} + 3d^{[n]}),\\
c^{[n+1]} &= c^{[n]} + \beta(b^{[n]} + 2e^{[n]}) + \beta^2 d^{[n]} + \left(\beta^2 - 4\beta\mathfrak{c}_6 + \frac{\mathfrak{c}_5^2}{4}\right)\left(a^{[n]} + d^{[n]}\right),
\end{aligned}$$

with the initial conditions $a^{[0]} = a$, $b^{[0]} = b$, $c^{[0]} = c$, $d^{[0]} = d$, and $e^{[0]} = e$. The solution of this system is

$$\begin{aligned}
&a^{[n]} = a, \quad b^{[n]} = b + 6\beta n(an + d), \quad d^{[n]} = 2an + d,\\
&e^{[n]} = bn + e + 2d\beta n^2 + \beta n^2(2an + d),\\
&c^{[n]} = \phi(\beta n^2) + 2\beta n\psi(\beta n^2) - n\left(4\beta\mathfrak{c}_6 - \frac{\mathfrak{c}_5^2}{4}\right)(an + d).
\end{aligned}$$

Thus, applying a limiting process (as $q \to 1$) on Theorem 9.2, we may infer and then prove rigorously the following theorem.

Theorem 9.3. *Consider the lattice*

$$x(s) = \mathfrak{c}_4 s^2 + \mathfrak{c}_5 s + \mathfrak{c}_6.$$

Suppose that $\mathbf{u} \in \mathcal{P}'$ *satisfies the functional differential* (9.2)*, with* ϕ *and* ψ *given by* (7.2)*, where at least one of these polynomials is nonzero. Then* $\mathbf{u}$ *is regular if and only if*

$$d_n \neq 0, \quad \phi^{[n]}\left(-\beta n^2 - \frac{e_n}{d_{2n}}\right) \neq 0, \quad n \in \mathbb{N}_0, \tag{9.91}$$

where $d_n := an + d$, $e_n := bn + e + 2\beta d n^2$, *and*

$$\phi^{[n]}(z) = az^2 + (b + 6\beta n d_n)z + \phi(\beta n^2) + 2\beta n \psi(\beta n^2) - \frac{n}{4}\left(16\beta \mathfrak{c}_6 - \mathfrak{c}_5^2\right) d_n.$$

Moreover, under these conditions, the monic OPS $\{P_n\}_{n\geq 0}$ *with respect to* $\mathbf{u}$ *satisfies* (9.81) *with*

$$B_n = \frac{ne_{n-1}}{d_{2n-2}} - \frac{(n+1)e_n}{d_{2n}} - 2\beta n(n-1), \tag{9.92}$$

$$C_{n+1} = -\frac{(n+1)d_{n-1}}{d_{2n-1}d_{2n+1}}\phi^{[n]}\left(-\beta n^2 - \frac{e_n}{d_{2n}}\right). \tag{9.93}$$

In addition, the following functional Rodrigues formula holds:

$$P_n \mathbf{u} = k_n \mathbf{D}^n \mathbf{u}^{[n]}, \quad k_n := (-1)^n \prod_{j=1}^{n} d_{n+j-2}^{-1}. \tag{9.94}$$

Remark 9.2. *For* $x(s) = \mathfrak{c}_6$ *we immediately recover Theorem 7.2, see Remark 1.1.*

9.3 Two Examples: Racah and Askey-Wilson Polynomials

In this section we derive the recurrence coefficients for the Racah and the Askey-Wilson polynomials [23, 24], using as departure point their associated functional equations. In this case, an important open problem is to obtain an analog of the Theorem 8.2.

9.3.1 The Racah Polynomials

Consider the quadratic lattice

$$x(s) = s(s + a + b + 1),$$

where $a, b, c, d \in \mathbb{C}$. Let $\mathbf{u} \in \mathcal{P}'$ be a functional satisfying (9.2), where

$$\begin{aligned}\phi(z) &:= 2z^2 + ((a + b + 2c + 3)d + c(a - b + 3) + 2(a + b + ab + 2))z \\ &\quad + (1 + a)(1 + d)(a + b + 1)(b + c + 1), \\ \psi(z) &:= 2(d + c + 2)z + 2(1 + a)(1 + d)(b + c + 1).\end{aligned}$$

According to (9.91) in Theorem 9.3, $\mathbf{u}$ is regular if and only if

$$\{a, c, d, d + c - 1, b + c, c + d - a, d - b\} \cap \mathbb{Z}^- = \emptyset.$$

Under these conditions, by (9.92)–(9.93), the recurrence coefficients for the monic OPS $\{P_n\}_{n\geq 0}$ with respect to $\mathbf{u}$ are given by

$$\begin{aligned}B_n &= -\frac{(n + a + 1)(n + d + 1)(n + b + c + 1)(n + d + c + 1)}{(2n + d + c + 1)(2n + d + c + 2)} \\ &\quad - \frac{n(n + c)(n + d + c - a)(n + d - b)}{(2n + d + c)(2n + d + c + 1)}, \\ C_{n+1} &= (n + 1)(n + a + 1)(n + c + 1)(n + d + 1) \\ &\quad \times \frac{(n + d + c + 1)(n + b + c + 1)(n + c + d - a + 1)(n + d - b + 1)}{(2n + d + c + 1)(2n + d + c + 2)^2(2n + d + c + 3)}.\end{aligned}$$

Therefore,

$$P_n(z) = R_n(z; d, c, a, b),$$

$\{R_n(.; d, c, a, b)\}_{n\geq 0}$ being the sequence of the monic Racah polynomials, see [24, Section 9.2].

9.3.2 The Askey-Wilson Polynomials

Consider the q-quadratic lattice

$$x(s) = \mathfrak{c}_1 q^{-s} + \mathfrak{c}_2 q^s + \mathfrak{c}_3.$$

Let $\mathbf{u}$ be a linear functional on $\mathcal{P}$ satisfying (9.2), where ϕ and ψ are given by

$$\begin{aligned}\phi(z) &= 2(1+abcd)(z-\mathfrak{c}_3)^2 \\ &\quad - 2\sqrt{\mathfrak{c}_1\mathfrak{c}_2}(a+b+c+d+abc+abd+acd+bcd)(z-\mathfrak{c}_3) \\ &\quad + 4\mathfrak{c}_1\mathfrak{c}_2(ab+ac+ad+bc+bd+cd-abcd-1), \\ \psi(z) &= \frac{4q^{1/2}}{q-1}\big((abcd-1)(z-\mathfrak{c}_3) \\ &\quad +\sqrt{\mathfrak{c}_1\mathfrak{c}_2}(a+b+c+d-abc-abd-acd-bcd)\big),\end{aligned}$$

with $a,b,c,d \in \mathbb{C}$. The parameter d_n given by (9.44) reads as

$$d_n = -\left(\frac{q^{1/2}-q^{-1/2}}{2}\right)^{-1} q^{-n/2}\left(1-abcdq^n\right).$$

By Theorem 9.2, $\mathbf{u}$ is regular if and only if

$$\begin{aligned}&\mathfrak{c}_1\mathfrak{c}_2(1-abcdq^n)(1-abq^n)(1-acq^n) \\ &\qquad \times(1-adq^n)(1-bcq^n)(1-bdq^n)(1-cdq^n) \neq 0, \quad n \in \mathbb{N}_0.\end{aligned}$$

Under these conditions, using formulas (9.82)–(9.83) in Theorem 9.2, we obtain the recurrence coefficients for the monic OPS $\{P_n\}_{n\geq 0}$ with respect to $\mathbf{u}$:

$$\begin{aligned}B_n &= \mathfrak{c}_3 + 2\sqrt{\mathfrak{c}_1\mathfrak{c}_2}\left(a+\frac{1}{a}-\frac{(1-abq^n)(1-acq^n)(1-adq^n)(1-abcdq^{n-1})}{a(1-abcdq^{2n-1})(1-abcdq^{2n})}\right. \\ &\quad \left. -\frac{a(1-q^n)(1-bcq^{n-1})(1-bdq^{n-1})(1-cdq^{n-1})}{(1-abcdq^{2n-1})(1-abcdq^{2n-2})}\right)\end{aligned}$$

(if $a=0$, define B_n by continuity, taking $a \to 0$ in the preceding expression), and

$$\begin{aligned}C_{n+1} &= \mathfrak{c}_1\mathfrak{c}_2(1-q^{n+1})(1-abcdq^{n-1}) \\ &\quad \times \frac{(1-abq^n)(1-acq^n)(1-adq^n)(1-bcq^n)(1-bdq^n)(1-cdq^n)}{(1-abcdq^{2n-1})(1-abcdq^{2n})^2(1-abcdq^{2n+1})}.\end{aligned}$$

Hence

$$P_n(z) = 2^n(\mathfrak{c}_1\mathfrak{c}_2)^{n/2}Q_n\left(\frac{z-\mathfrak{c}_3}{2\sqrt{\mathfrak{c}_1\mathfrak{c}_2}};a,b,c,d \mid q\right),$$

where $\{Q_n(\cdot;a,b,c,d|q)\}_{n\geq 0}$ are the monic Askey-Wilson polynomials (see, for instance, [24, Section 14.1]).

Exercises

1. Prove Lemma 9.1.

2. Prove Lemma 9.3.

Chapter 10

Classical Orthogonal Polynomials: The Positive Definite Case

In this chapter we analyze the classical functionals in the positive definite case. To be more precise, we find the conditions ensuring that the classical functionals are positive definite, and then, under such conditions, we show that these functionals may be represented uniquely by simple weight functions (via proper or improper Riemann integrals). Of course, up to the affine changes of the variables, we only need to analyze the positive definiteness of the canonical forms described in Table 8.1.

10.1 The Positive Definite Case

We begin by stating the following elementary proposition.

Lemma 10.1. *Let (ξ, η) be a bounded or unbounded interval of real numbers. Let $\omega : (\xi, \eta) \to \mathbb{R}$ be a function fulfilling the following four properties:*

(i) $\omega \in C^1(\xi, \eta)$ *and* $\omega(x) > 0$ *for each* $x \in (\xi, \eta)$;

(ii)
$$\int_\xi^\eta |x|^k \omega(x)\,\mathrm{d}x < \infty, \quad k \in \mathbb{N}_0;$$

(iii) *there exist real polynomials ϕ and ψ such that ω fulfills the first-order ordinary differential equation*
$$(\phi\omega)' = \psi\omega \text{ on } (\xi, \eta); \tag{10.1}$$

(iv)
$$\lim_{x \to \xi^+} x^k \phi(x)\omega(x) = \lim_{x \to \eta^-} x^k \phi(x)\omega(x) = 0, \quad k \in \mathbb{N}_0.$$

DOI: 10.1201/9781003432067-10

Define a functional $\mathbf{u}$ *on* $\mathcal{P}$ *by*

$$\langle \mathbf{u}, p \rangle := \int_{\xi}^{\eta} p(x)\omega(x)\,\mathrm{d}x, \quad p \in \mathcal{P}. \tag{10.2}$$

Then $\mathbf{u}$ *is a positive definite functional on* $[\xi, \eta]$*, and it fulfills the generalized functional differential equation*

$$D(\phi\mathbf{u}) = \psi\mathbf{u}. \tag{10.3}$$

Proof. The hypotheses (i)–(ii) ensure that $\mathbf{u}$ is well-defined. Take arbitrarily $p \in \mathcal{P}$ such that $p(x) \geq 0$ on $[\xi, \eta]$ and $p(x) \not\equiv 0$. Since p is continuous on $[\xi, \eta]$ and does not vanish identically there, then there exist $x_0 \in (\xi, \eta)$ and $\delta > 0$ so that $(x_0 - \delta, x_0 + \delta) \subset (\xi, \eta)$ and $p(x) > \epsilon := p(x_0)/2 > 0$ for each $x \in (x_0 - \delta, x_0 + \delta)$. Hence

$$\langle \mathbf{u}, p \rangle = \int_{\xi}^{\eta} p(x)\omega(x)\,\mathrm{d}x \geq \epsilon \int_{x_0-\delta}^{x_0+\delta} \omega(x)\,\mathrm{d}x > 0,$$

where the last equality follows from hypothesis (i). Thus $\mathbf{u}$ is positive definite on $[\xi, \eta]$. To prove that $\mathbf{u}$ satisfies (10.3), take $p \in \mathcal{P}$. Hence

$$\begin{aligned}
\langle D(\phi\mathbf{u}), p \rangle &= -\langle \mathbf{u}, \phi p' \rangle = -\int_{\xi}^{\eta} \phi p' \omega\,\mathrm{d}x \\
&= -\int_{\xi}^{\eta} \left((\phi\omega p)' - (\phi\omega)' p\right)\mathrm{d}x \\
&= \phi(x)\omega(x)p(x)\big|_{\xi}^{\eta} + \int_{\xi}^{\eta} \psi\omega p\,\mathrm{d}x = \langle \psi\mathbf{u}, p \rangle,
\end{aligned}$$

where we have used (iii) in the fourth equality, and (ii) and (iv) in the last one. □

Remark 10.1. *Under the conditions of Lemma 10.1, we say that* ω *is a weight function for* $\mathbf{u}$*, and that* $\mathbf{u}$ *is represented by the weight function* ω*; and we also say that the OPS with respect to* $\mathbf{u}$ *is orthogonal with respect to the weight function* ω*.*

10.1.1 Hermite Functional

From Table 8.2, the coefficients appearing in the three-term recurrence relation for the monic OPS with respect to the (canonical) Hermite functional, $\mathbf{u}_H$, satisfy

$$\beta_n = 0, \quad n \in \mathbb{N}_0; \quad \gamma_n = \frac{n}{2} > 0, \quad n \in \mathbb{N}.$$

Therefore, by the Favard theorem, $\mathbf{u}_H$ is positive definite. Next, we show that $\mathbf{u}_H$ is represented by a weight function, in the sense of (10.2). First, we take the polynomials ϕ and ψ from Table 8.1, so that

$$\phi(x) \equiv 1, \quad \psi(x) = -2x.$$

This gives the ordinary differential equation

$$\omega' = -2x\omega.$$

The general solution of this equation is Ce^{-x^2}, where C is an arbitrary real constant. Thus we choose the weight function

$$\omega(x) := e^{-x^2}, \quad x \in \mathbb{R}. \tag{10.4}$$

Notice that it is quite natural to take $(\xi, \eta) := \mathbb{R}$, since this is the largest interval where ω becomes positive, as required on hypothesis (i) appearing in Lemma 10.1. Of course, by construction, (iii) is also fulfilled. Moreover, one immediately sees that ω satisfies the remaining hypotheses (ii) and (iv). Thus, by Lemma 10.1 and Theorem 8.2,

$$\langle \mathbf{u}_H, p \rangle = \int_{-\infty}^{+\infty} p(x) e^{-x^2} \, \mathrm{d}x, \quad p \in \mathcal{P}, \tag{10.5}$$

meaning that $\mathbf{u}_H$ is represented by the weight function (10.4). The corresponding positive Borel measure is supported on $\mathbb{R}$, and the associated distribution function $\psi_H : \mathbb{R} \to \mathbb{R}$ is given by

$$\psi_H(x) := \int_{-\infty}^{x} e^{-t^2} \, \mathrm{d}t, \quad x \in \mathbb{R}. \tag{10.6}$$

Notice that $\mathbf{u}_H$ is uniquely determined by ψ_H. Indeed, this follows from the Corollary 4.1, by choosing $\theta = 2$ and hence noticing that

$$\int_{-\infty}^{+\infty} e^{\theta |x|} \, \mathrm{d}\psi_H(x) = 2 \int_{0}^{+\infty} e^{2x - x^2} \, \mathrm{d}x < \infty.$$

Finally, in accordance with the spectral theorem, we conclude that the Hermite polynomials are orthogonal in the positive definite sense with respect to a unique positive Borel measure supported on $\mathbb{R}$, and characterized by the distribution function (10.6). The reader would recognize here, up to normalization, the Gaussian (or normal) probability distribution function.

10.1.2 Laguerre Functional

From Table 8.2, the coefficients appearing in the three-term recurrence relation for the monic OPS with respect to the Laguerre functional, $\mathbf{u}_L^{(\alpha)}$, satisfy

$$\beta_n \in \mathbb{R}, \quad n \in \mathbb{N}_0 \Leftrightarrow \alpha \in \mathbb{R}; \quad \gamma_n > 0, \quad n \in \mathbb{N} \Leftrightarrow \alpha > -1.$$

Therefore, $\mathbf{u}_L^{(\alpha)}$ is positive definite if and only if $\alpha > -1$. To show that $\mathbf{u}_L^{(\alpha)}$ is represented by a weight function (if $\alpha > -1$), consider the corresponding polynomials

$$\phi(x) = x, \quad \psi(x) = -x + \alpha + 1,$$

given by Table 8.1. This gives the ordinary differential equation

$$(x\omega)' = (-x + \alpha + 1)\omega.$$

The general solution of this equation is $Cx^\alpha e^{-x}$, $C \in \mathbb{R}$. Thus we choose

$$\omega(x) := x^\alpha e^{-x}, \quad x \in (0, +\infty). \tag{10.7}$$

As before, we chose $(\xi, \eta) := (0, +\infty)$ since this is the largest interval where ω becomes positive. Thus, hypothesis (i) and (iii) in Lemma 10.1 are fulfilled. Moreover, since $\alpha > -1$, we have

$$\int_0^{+\infty} |x|^k w(x)\,\mathrm{d}x = \int_0^1 x^{\alpha+k} e^{-x}\,\mathrm{d}x + \int_1^{+\infty} x^{\alpha+k} e^{-x}\,\mathrm{d}x < \infty, \quad k \in \mathbb{N}_0.$$

The last two integrals converge. Indeed, on one hand,

$$\int_0^1 x^{\alpha+k} e^{-x}\,\mathrm{d}x \leq \int_0^1 x^{\alpha+k}\,\mathrm{d}x < \infty$$

(because $\alpha + k > -1$ for each $k \in \mathbb{N}_0$); on the other hand,

$$\int_1^{+\infty} x^{\alpha+k} e^{-x}\,\mathrm{d}x < \infty,$$

since $\int_1^{+\infty} x^{-s}\,\mathrm{d}x < \infty$ for an arbitrarily fixed $s > 1$, and

$$\frac{x^{\alpha+k} e^{-x}}{\frac{1}{x^s}} = x^{\alpha+k+s} e^{-x} \to 0 \quad (x \to +\infty).$$

Thus, ω satisfies hypothesis (ii). Of course, ω also satisfies (iv). Thus, by Lemma 10.1 and Theorem 8.2,

$$\langle \mathbf{u}_L^{(\alpha)}, p \rangle = \int_0^{+\infty} p(x) x^\alpha e^{-x}\,\mathrm{d}x, \quad p \in \mathcal{P}, \tag{10.8}$$

so $\mathbf{u}_L^{(\alpha)}$ is represented by the weight function (10.7). The corresponding positive Borel measure is supported on the closed interval $[0,+\infty)$, and the associated distribution function $\psi_L^{(\alpha)} : \mathbb{R} \to \mathbb{R}$ is given by

$$\psi_L^{(\alpha)}(x) := \int_{-\infty}^{x} t^\alpha e^{-t} \chi_{(0,+\infty)}(t)\, \mathrm{d}t, \quad x \in \mathbb{R}. \tag{10.9}$$

Notice that $\mathbf{u}_L^{(\alpha)}$ is uniquely determined by $\psi_L^{(\alpha)}$. This follows, e.g., from Corollary 4.1, by choosing $0 < \theta < 1$ and hence noticing that

$$\int_{-\infty}^{+\infty} e^{\theta|x|}\, \mathrm{d}\psi_L^{(\alpha)}(x) = \int_0^1 x^\alpha e^{(\theta-1)x}\, \mathrm{d}x + \int_1^{+\infty} x^\alpha e^{(\theta-1)x}\, \mathrm{d}x < \infty.$$

Finally, in accordance with the spectral theorem, we conclude that if $\alpha > -1$ the Laguerre polynomials are orthogonal in the positive definite sense with respect to a unique positive Borel measure supported on $[0,+\infty)$, and characterized by the distribution function (10.9). The reader would recognize here, up to normalization, the gamma probability distribution function.

10.1.3 Jacobi Functional

By Table 8.2, the coefficients appearing in the three-term recurrence relation for the monic OPS with respect to the Jacobi functional, $\mathbf{u}_J^{(\alpha,\beta)}$, satisfy

$$\beta_n \in \mathbb{R}, \quad n \in \mathbb{N}_0 \;\Leftrightarrow\; \alpha,\beta \in \mathbb{R}; \quad \gamma_n > 0, \quad n \in \mathbb{N} \;\Leftrightarrow\; \alpha,\beta > -1.$$

Therefore, $\mathbf{u}_J^{(\alpha,\beta)}$ is positive definite if and only if $\alpha > -1$ and $\beta > -1$. To show that $\mathbf{u}_J^{(\alpha,\beta)}$ is represented by a weight function (if $\alpha > -1$ and $\beta > -1$), consider the corresponding polynomials

$$\phi(x) = 1 - x^2 \quad \psi(x) = -(\alpha+\beta+2)x + \beta - \alpha,$$

given by Table 8.1. This gives the ordinary differential equation

$$\big((1-x^2)\omega\big)' = \big(-(\alpha+\beta+2)x + \beta - \alpha\big)\omega.$$

The general solution of this equation is $C(1-x)^\alpha(1+x)^\beta$, $C \in \mathbb{R}$. Thus we choose

$$\omega(x) := (1-x)^\alpha(1+x)^\beta, \quad x \in (-1,1). \tag{10.10}$$

Clearly, the hypotheses (i), (iii), and (iv) appearing in Lemma 10.1 are fulfilled, where $(\xi,\eta) := (-1,1)$. Moreover, since $\alpha > -1$ and $\beta > -1$, we have, for each $k \in \mathbb{N}_0$,

$$\int_{-1}^{1} |x|^k w(x)\, \mathrm{d}x \le 2^\alpha \int_{-1}^{0} (1+x)^\beta\, \mathrm{d}x + 2^\beta \int_0^1 (1-x)^\alpha\, \mathrm{d}x < \infty,$$

TABLE 10.1: Weight functions for the canonical forms.

$\mathbf{u}$	Interval of orthogonality	$w(x)$	Restrictions
$\mathbf{u}_J^{(\alpha,\beta)}$	$[-1,1]$	$(1-x)^\alpha(1+x)^\beta$	$\alpha>-1,\ \beta>-1$
$\mathbf{u}_L^{(\alpha)}$	$[0,+\infty)$	$x^\alpha e^{-x}$	$\alpha>-1$
$\mathbf{u}_H$	$(-\infty,+\infty)$	e^{-x^2}	—

and so ω satisfies hypothesis (ii). Thus, by Lemma 10.1 and Theorem 8.2,

$$\langle \mathbf{u}_J^{(\alpha,\beta)}, p\rangle = \int_{-1}^{1} p(x)(1-x)^\alpha(1+x)^\beta\, dx, \quad p\in\mathcal{P}, \tag{10.11}$$

hence $\mathbf{u}_J^{(\alpha,\beta)}$ is represented by the weight function (10.10). The corresponding positive Borel measure is supported on the closed interval $[-1,1]$, and the associated distribution function $\psi_J^{(\alpha,\beta)}:\mathbb{R}\to\mathbb{R}$ is given by

$$\psi_J^{(\alpha,\beta)}(x) := \int_{-\infty}^{x} (1-t)^\alpha(1+t)^\beta \chi_{(-1,1)}(t)\, dt, \quad x\in\mathbb{R}. \tag{10.12}$$

Notice that, since the sequences $\{\beta_n\}_{n\geq 0}$ and $\{\gamma_n\}_{n\geq 1}$ are bounded, then, by Theorem 4.7, $\mathbf{u}_J^{(\alpha,\beta)}$ is uniquely determined by $\psi_J^{(\alpha,\beta)}$. Therefore, we conclude that if $\alpha>-1$ and $\beta>-1$ then the Jacobi polynomials are orthogonal in the positive definite sense with respect to a unique positive Borel measure supported on $[-1,1]$, and characterized by the distribution function (10.12). The reader would recognize here, up to normalization, the beta probability distribution function.

10.1.4 Bessel Functional

Consider the coefficients given by Table 8.2 for the three-term recurrence relation of the monic OPS with respect to the Bessel functional, $\mathbf{u}_B^{(\alpha)}$. We see that the condition $\alpha\in\mathbb{R}\setminus\{-2,-3,-4,\dots\}$ is necessary for $\beta_n\in\mathbb{R}$ for all $n\in\mathbb{N}_0$. Under this condition, we see that $\gamma_n>0$ for all $n\in\mathbb{N}$ if and only if α fulfills the property

$$\alpha<-(2n+1) \quad \text{or} \quad -(2n-1)<\alpha<-n, \quad n\in\mathbb{N}.$$

Clearly, there is no α fulfilling this property. Therefore, $\mathbf{u}_B^{(\alpha)}$ is not positive definite independently of the choice of the parameter α.

10.2 Orthogonality of the Bessel Polynomials

We have seen that the Bessel OPS $\{\widehat{B}_n^{(\alpha)}\}_{n\geq 0}$ is not an OPS with respect to a positive definite functional. Despite this fact, Krall and Frink [27] proved that $\{\widehat{B}_n^{(\alpha)}\}_{n\geq 0}$ fulfills the orthogonality relations (10.15) below, where the integration is over the unit circle $\mathbb{S}^1 := \{z \in \mathbb{C} : |z| = 1\}$ (or any closed contour around the origin) and the "weight" function is given by

$$\rho^{(\alpha)}(z) := \frac{1}{2\pi i}\left(1+\alpha+\sum_{k=1}^{\infty}\frac{1}{(\alpha+2)_{k-1}}\left(-\frac{2}{z}\right)^k\right), \quad z\in\mathbb{C}\setminus\{0\}. \quad (10.13)$$

The ratio test ensures that the series in (10.13) converges absolutely on $\mathbb{C}\setminus\{0\}$ and uniformly on each compact subset of this set. This function $\rho\equiv\rho^{(\alpha)}(z)$ fulfills

$$(z^2\rho)' = ((\alpha+2)z+2)\rho - \frac{\alpha(\alpha+1)}{2\pi i}z, \quad z\in\mathbb{C}\setminus\{0\}. \quad (10.14)$$

Indeed, since

$$\rho'(z) = \frac{1}{4\pi i}\sum_{k=0}^{\infty}\frac{k+1}{(\alpha+2)_k}\left(-\frac{2}{z}\right)^{k+2},$$

we deduce

$$\begin{aligned}
&(z^2\rho)'(z) - ((\alpha+2)z+2)\,\rho(z) = z^2\rho'(z) - (\alpha z+2)\rho(z)\\
&= \frac{1}{\pi i}\sum_{k=0}^{\infty}\frac{k+1}{(\alpha+2)_k}\left(-\frac{2}{z}\right)^k\\
&\quad - \frac{\alpha z+2}{2\pi i}\left(\alpha+1-\frac{2}{z}\sum_{k=0}^{\infty}\frac{1}{(\alpha+2)_k}\left(-\frac{2}{z}\right)^k\right)\\
&= -\frac{\alpha(\alpha+1)}{2\pi i}z + \frac{1}{\pi i}\sum_{k=0}^{\infty}\frac{k+1+\alpha}{(\alpha+2)_k}\left(-\frac{2}{z}\right)^k\\
&\quad - \frac{1}{\pi i}\left(\alpha+1-\frac{2}{z}\sum_{k=0}^{\infty}\frac{1}{(\alpha+2)_k}\left(-\frac{2}{z}\right)^k\right) = -\frac{\alpha(\alpha+1)}{2\pi i}z.
\end{aligned}$$

Theorem 10.1. *Let* $\{\widehat{B}_n^{(\alpha)}\}_{n\geq 0}$, $\alpha \in \mathbb{C}\setminus\{-2,-3,-4,\dots\}$, *be the Bessel monic OPS, and let* $\rho^{(\alpha)}$ *be defined as in* (10.13). *Then*

$$\int_{\mathbb{S}^1} \widehat{B}_m^{(\alpha)}(z)\widehat{B}_n^{(\alpha)}(z)\rho^{(\alpha)}(z)\,\mathrm{d}z = \frac{2^{2n+1}(-1)^{n+1}n!}{(\alpha+2)_{2n}(n+\alpha+1)_n}\delta_{m,n}, \quad m\in\mathbb{N}_0. \tag{10.15}$$

Proof. From C5 in Theorem 8.1 and Table 8.1, $y_n := \widehat{B}_n^{(\alpha)}(z)$ fulfills

$$z^2y_n'' + ((\alpha+2)z+2)y_n' = n(n+\alpha+1)y_n, \quad n\in\mathbb{N}. \tag{10.16}$$

Multiplying both sides of (10.16) by ρ and taking into account (10.14), we deduce

$$(z^2\rho y_n')' + \frac{\alpha(\alpha+1)}{2\pi i}zy_n' = n(n+\alpha+1)\rho y_n.$$

Multiplying both sides of this equality by y_m and then integrating around $\mathbb{S}^1$,

$$\int_{\mathbb{S}^1}(z^2\rho y_n')'y_m\,\mathrm{d}z + \frac{\alpha(\alpha+1)}{2\pi i}\int_{\mathbb{S}^1} zy_n'y_m\,\mathrm{d}z = n(n+\alpha+1)\int_{\mathbb{S}^1} y_ny_m\rho\,\mathrm{d}z.$$

Clearly, by Cauchy's theorem,

$$\int_{\mathbb{S}^1} zy_n'y_m\,\mathrm{d}z = 0.$$

Moreover, integrating by parts[1], we deduce

$$\int_{\mathbb{S}^1}(z^2\rho y_n')'y_m\,\mathrm{d}z = -\int_{\mathbb{S}^1} z^2\rho y_n'y_m'\,\mathrm{d}z.$$

Hence the above equality reduces to

$$n(n+\alpha+1)\int_{\mathbb{S}^1} y_ny_m\rho\,\mathrm{d}z = -\int_{\mathbb{S}^1} z^2y_n'y_m'\rho\,\mathrm{d}z. \tag{10.17}$$

Interchanging n and m and subtracting the resulting equality to (10.17), yields

$$(n-m)(n+m+\alpha+1)\int_{\mathbb{S}^1} y_ny_m\rho\,\mathrm{d}z = 0. \tag{10.18}$$

[1] Recall that if f and g are complex functions holomorphic on a neighborhood of the image of a differentiable and closed path γ, then

$$\int_\gamma f'(z)g(z)\,\mathrm{d}z = -\int_\gamma f(z)g'(z)\,\mathrm{d}z.$$

Since $-(\alpha+1) \notin \mathbb{N}_0$ then $n+m+\alpha+1 \neq 0$ if $n \neq m$. Thus (10.18) gives us

$$\int_{\mathbb{S}^1} y_n y_m \rho \, \mathrm{d}z = 0 \quad \text{if} \quad n \neq m. \tag{10.19}$$

This proves (10.15) whenever $n \neq m$. If $n = m$, from (10.17) we find

$$I_n := \int_{\mathbb{S}^1} y_n^2 \rho \, \mathrm{d}z = -\frac{1}{n(n+\alpha+1)} \int_{\mathbb{S}^1} z^2 y_n' \cdot y_n' \rho \, \mathrm{d}z, \quad n \in \mathbb{N}. \tag{10.20}$$

By C2 in Theorem 8.1, we have $z^2 y_n' = a_n y_{n+1} + b_n y_n + c_n y_{n-1}$. Moreover, clearly,

$$y_n' = n y_{n-1} + \sum_{j=0}^{n-2} a_{nj} y_j$$

for some complex numbers a_{nj}. Substituting these expressions into the integrand on the right-hand side of (10.20) and using (10.19), we obtain

$$I_n := \int_{\mathbb{S}^1} y_n^2 \rho \, \mathrm{d}z = -\frac{c_n}{n+\alpha+1} \int_{\mathbb{S}^1} y_{n-1}^2 \rho \, \mathrm{d}z = \gamma_n I_{n-1}, \quad n \in \mathbb{N}, \tag{10.21}$$

where the last equality holds since $c_n = -d_{n-1}\gamma_n = -(n+\alpha+1)\gamma_n$ (see Theorem 8.1 and Table 8.1). Iterating (10.21) we deduce $I_n = \gamma_n \gamma_{n-1} \cdots \gamma_1 I_0$, hence using the expression for γ_n given in Table 8.2, we deduce

$$I_n = \frac{(-4)^n n!}{(\alpha+2)_{2n}(n+\alpha+1)_n} I_0, \quad n \in \mathbb{N}_0. \tag{10.22}$$

It remains to compute I_0. Since ρ is given by the Laurent series (10.13), one see by the definition of residue that $\operatorname{Res}(\rho; z=0) = -1(\pi i)$, hence, by the residue theorem,

$$I_0 := \int_{\mathbb{S}^1} \rho \, \mathrm{d}z = 2\pi i \operatorname{Res}(\rho; z=0) = -2. \tag{10.23}$$

Inserting (10.23) into (10.22) yields (10.15) for $n = m$. This completes the proof. □

Remark 10.2. *The function* $\omega_\alpha(z) := z^\alpha e^{-2/z}$ *fulfills the differential equation*

$$\left(z^2 \omega_\alpha(z)\right)' = \left((\alpha+2)z + 2\right)\omega_\alpha(z), \quad z \in \mathbb{C} \setminus \{0\}. \tag{10.24}$$

This suggests using ω_α *instead of* $\rho^{(\alpha)}$ *as "weight" function in the orthogonality relations (10.15). However,* ω_α *is a multivalued function if* α *is not an integer number and this may be inconvenient for integration around the origin 0. In general,* $\rho^{(\alpha)}$ *and* ω_α *yield different orthogonality relations for the Bessel polynomials, unless* $\alpha = 0$ *or* $\alpha = -1$ *(compare equations (10.14) and (10.24), and see Exercise 4).*

10.3 Explicit Expressions for the Classical Orthogonal Polynomials

10.3.1 The Rodrigues Formula

Theorem 7.2 states that the monic OPS $\{P_n\}_{n\geq 0}$ with respect to a classical functional $\mathbf{u}$ (which does not need to be positive definite) satisfies a distributional Rodrigues formula, involving P_n and $\mathbf{u}$. In the next theorem we prove that if, in addition, $\mathbf{u}$ is (classical and) represented by the weight function ω, then also a Rodrigues formula involving P_n and ω hods.

Theorem 10.2 (Rodrigues formula). *Assume the hypothesis of Lemma 10.1, so that* $\mathbf{u}$ *is positive definite and represented by the weight function* ω*, as in* (10.2). *Assume further that* $\omega \in C^\infty(\xi,\eta)$*, and* ϕ *and* ψ *are nonzero real polynomials,* $\phi \in \mathcal{P}_2$ *and* $\psi \in \mathcal{P}_1$. *Let* $\{P_n\}_{n\geq 0}$ *be the monic OPS with respect to* $\mathbf{u}$. *Then, for each* $n \in \mathbb{N}_0$,

$$P_n(x) = \frac{k_n}{w(x)}\frac{\mathrm{d}^n}{\mathrm{d}x^n}\big(\phi^n(x)\omega(x)\big), \quad \xi < x < \eta, \tag{10.25}$$

where

$$k_n := \prod_{i=0}^{n-1}\frac{1}{d_{n+i-1}}, \quad d_k := \frac{k}{2}\phi'' + \psi'.$$

Proof. By Lemma 10.1, $\mathbf{u}$ fulfills the functional equation

$$D(\phi\mathbf{u}) = \psi\mathbf{u}. \tag{10.26}$$

Therefore, since $\mathbf{u}$ is regular (because it is positive definite), Theorem 7.2 ensures that $d_k \neq 0$ for all $k \in \mathbb{N}_0$, and the functional Rodrigues formula holds:

$$P_n\mathbf{u} = k_n D^n(\phi^n\mathbf{u}). \tag{10.27}$$

While proving Theorem 7.1, and up to normalization (being $P_n = k_n R_n$), we have deduced (10.27) from (10.26) by purely algebraic arguments, the essential tool in the proof being the functional Leibnitz formula for the derivative of order n of the functional $\phi\mathbf{u}$ (the left product of the functional $\mathbf{u}$ by the polynomial ϕ). Therefore, since the weight function ω fulfills the ordinary differential equation (which can be regarded as an analog version of ordinary functions of the functional differential equation (10.26))

$$(\phi\omega)' = \psi\omega \quad \text{on} \quad (\xi,\eta), \tag{10.28}$$

we see *mutatis mutandis* that the steps of the proof of Theorem 7.1 may be followed (replacing therein $\mathbf{u}$ by ω, and $\mathbf{u}^{[n]}$ by $\omega^{[n]} := \phi^n\omega$, and considering the ordinary derivative instead of the functional derivative) allowing us to deduce (10.25) from (10.28) and the Leibnitz formula for the product of ordinary functions. □

The Rodrigues formula (10.25) gives an explicit representation of P_n as a derivative of order n of a simple real function, divided by the weight function. This representation is very useful in many areas, e.g., in Number Theory, or in Physics. In the next sections we use (10.25) to derive explicit expressions for P_n as a linear combination of powers of x. Again, up to an affine change of variables, we can restrict our study to the canonical forms described in the previous sections.

10.3.2 Jacobi Polynomials

Consider $\alpha > -1$ and $\beta > -1$. Substituting in (10.25) the explicit expression of ϕ, ψ, k_n, and ω appearing in Tables 8.1, 8.3, and 10.1, we may write, for each $n \in \mathbb{N}_0$ and $-1 < x < 1$,

$$\widehat{P}_n^{(\alpha,\beta)}(x) = \frac{(-1)^n}{(n+\alpha+\beta+1)_n}\frac{1}{(1-x)^\alpha(1+x)^\beta}\frac{\mathrm{d}^n}{\mathrm{d}x^n}\left((1-x)^{n+\alpha}(1+x)^{n+\beta}\right).$$

By the Leibniz formula for the nth derivative of a product, and making use of the generalized binomial coefficient, defined by

$$\binom{z}{0} := 1, \quad \binom{z}{k} := \frac{z(z-1)\cdots(z-k+1)}{k!}, \qquad z \in \mathbb{C}, \quad k \in \mathbb{N},$$

we deduce

$$\widehat{P}_n^{(\alpha,\beta)}(x) = \frac{1}{\binom{2n+\alpha+\beta}{n}}\sum_{k=0}^{n}\binom{n+\alpha}{n-k}\binom{n+\beta}{k}(x-1)^k(x+1)^{n-k}. \quad (10.29)$$

In the literature on OP it is usual to consider (nonmonic) Jacobi polynomials $\{P_n^{(\alpha,\beta)}\}_{n\geq 0}$ normalized so that

$$P_n^{(\alpha,\beta)}(1) = \binom{n+\alpha}{n}, \quad (10.30)$$

called the standard normalization for Jacobi polynomials. Therefore, since, by (10.29),

$$\widehat{P}_n^{(\alpha,\beta)}(1) = \frac{2^n}{\binom{2n+\alpha+\beta}{n}}\binom{n+\alpha}{n} = \frac{2^n(\alpha+1)_n}{(n+\alpha+\beta+1)_n}, \quad (10.31)$$

we conclude that the relation between the Jacobi polynomials with standard normalization and the monic Jacobi polynomials is

$$P_n^{(\alpha,\beta)}(x) = 2^{-n}\binom{2n+\alpha+\beta}{n}\widehat{P}_n^{(\alpha,\beta)}(x). \tag{10.32}$$

This together with (10.29) leads to the explicit expression of the Jacobi polynomials with the standard normalization (10.32):

$$P_n^{(\alpha,\beta)}(x) = 2^{-n}\sum_{k=0}^{n}\binom{n+\alpha}{n-k}\binom{n+\beta}{k}(x-1)^k(x+1)^{n-k}.$$

We also point out the following useful relation:

$$P_n^{(\alpha,\beta)}(-x) = (-1)^n P_n^{(\beta,\alpha)}(x). \tag{10.33}$$

(Clearly this relation holds also for the monic polynomials.) Also, from (10.30), (10.31), and (10.33),

$$P_n^{(\alpha,\beta)}(-1) = (-1)^n\binom{n+\beta}{n}, \quad \widehat{P}_n^{(\alpha,\beta)}(-1) = \frac{(-2)^n(\beta+1)_n}{(n+\alpha+\beta+1)_n}.$$

Finally, for each $k \in \mathbb{N}_0$, the following formula holds:

$$\frac{\mathrm{d}^k}{\mathrm{d}x^k}\left(\widehat{P}_n^{(\alpha,\beta)}(x)\right) = (n-k+1)_k\,\widehat{P}_{n-k}^{(\alpha+k,\beta+k)}(x), \quad n \geq k. \tag{10.34}$$

10.3.3 Laguerre Polynomials

Let $\alpha > -1$. Substituting in (10.25) the explicit expression of ϕ, ψ, k_n, and ω appearing in Tables 8.1, 8.3, and 10.1, we may write, for each $n \in \mathbb{N}_0$ and $x > 0$,

$$\widehat{L}_n^{(\alpha)}(x) = (-1)^n x^{-\alpha}e^x\frac{\mathrm{d}^n}{\mathrm{d}x^n}\left(x^{n+\alpha}e^{-x}\right).$$

By the Leibniz formula for the nth derivative of a product, we deduce

$$\widehat{L}_n^{(\alpha)}(x) = (-1)^n n!\sum_{k=0}^{n}\binom{n+\alpha}{n-k}\frac{(-x)^k}{k!}. \tag{10.35}$$

Considering Laguerre polynomials $\{L_n^{(\alpha)}\}_{n\geq 0}$ with standard normalization, i.e.,

$$L_n^{(\alpha)}(0) = \binom{n+\alpha}{n},$$

then, since, by (10.35),

$$\widehat{L}_n^{(\alpha)}(0) = (-1)^n n!\binom{n+\alpha}{n} = (-1)^n(\alpha+1)_n,$$

we see that the relation between the Laguerre polynomials with standard normalization and the monic Laguerre polynomials is

$$L_n^{(\alpha)}(x) = \frac{(-1)^n}{n!}\,\widehat{L}_n^{(\alpha)}(x).$$

This together with (10.35) leads to the explicit expression of the Laguerre polynomials with the standard normalization:

$$L_n^{(\alpha)}(x) = \sum_{k=0}^{n} \binom{n+\alpha}{n-k} \frac{(-x)^k}{k!}.$$

We also point out that, for each $k \in \mathbb{N}_0$, the following formula holds:

$$\frac{\mathrm{d}^k}{\mathrm{d}x^k}\left(\widehat{L}_n^{(\alpha)}(x)\right) = (n-k+1)_k\,\widehat{L}_{n-k}^{(\alpha+k)}(x), \quad n \geq k. \tag{10.36}$$

10.3.4 Hermite Polynomials

Substituting in (10.25) the explicit expression of ϕ, ψ, k_n, and ω appearing in Tables 8.1, 8.3, and 10.1, we may write, for each $n \in \mathbb{N}_0$ and $x \in \mathbb{R}$,

$$\widehat{H}_n(x) = \frac{(-1)^n}{2^n}\, e^{x^2} \frac{\mathrm{d}^n}{\mathrm{d}x^n}\{e^{-x^2}\}.$$

Using this formula we can derive the explicit expression for the Hermite polynomials. Nevertheless, we will obtain such a formula in a different way. By Lemma 7.4, for each $k \in \mathbb{N}_0$,

$$\left\{\widehat{H}_n^{[k]} := \frac{1}{(n+1)_k}\frac{\mathrm{d}^k \widehat{H}_{n+k}}{\mathrm{d}x^k}\right\}_{n\geq 0}$$

is a monic OPS with respect to the functional $\mathbf{u}^{[k]} := \phi^k \mathbf{u}_H = \mathbf{u}_H$ (since $\phi \equiv 1$), hence $\widehat{H}_n^{[k]} \equiv H_n$, and so

$$\frac{\mathrm{d}^k}{\mathrm{d}x^k}\{\widehat{H}_n(x)\} = (n-k+1)_k\,\widehat{H}_{n-k}(x), \quad n \geq k. \tag{10.37}$$

Therefore, using the McLaurin formula, and since $(n-k+1)_k = k!\binom{n}{k}$, we may write

$$\widehat{H}_n(x) = \sum_{k=0}^{n} \frac{\frac{\mathrm{d}^k \widehat{H}_n}{\mathrm{d}x^k}(0)}{k!}\, x^k = \sum_{k=0}^{n} \binom{n}{k} \widehat{H}_{n-k}(0)\, x^k$$

$$= \sum_{j=0}^{n} \binom{n}{j} \widehat{H}_j(0)\, x^{n-j}. \tag{10.38}$$

To compute $\widehat{H}_j(0)$, we start with the three-term recurrence relation for $\{\widehat{H}_n\}_{n\geq 0}$ (see Table 8.2):

$$\widehat{H}_{n+1}(x) = x\,\widehat{H}_n(x) - \frac{n}{2}\widehat{H}_{n-1}(x)$$

with $\widehat{H}_{-1} = 0$ and $\widehat{H}_0 = 1$. Thus $\widehat{H}_{n+1}(0) = -\frac{n}{2}\,\widehat{H}_{n-1}(0)$, hence

$$\widehat{H}_{2n-1}(0) = 0, \quad \widehat{H}_{2n}(0) = (-1)^n \frac{(2n-1)!!}{2^n}. \tag{10.39}$$

Inserting (10.39) into (10.38) we easily deduce the desired explicit expression:

$$\widehat{H}_n(x) = \frac{n!}{2^n} \sum_{k=0}^{\lfloor n/2 \rfloor} \frac{(-1)^k (2x)^{n-2k}}{(n-2k)!k!}. \tag{10.40}$$

The standard normalization for the Hermite polynomials is $\{H_n\}_{n\geq 0}$ given by

$$H_n(x) = 2^n \widehat{H}_n(x) = n! \sum_{k=0}^{\lfloor n/2 \rfloor} \frac{(-1)^k (2x)^{n-2k}}{(n-2k)!k!}. \tag{10.41}$$

10.3.5 Bessel Polynomials

The Bessel functional $\mathbf{u}_B^{(\alpha)}$ fulfills the functional differential equation $D(\phi \mathbf{u}_B^{(\alpha)}) = \psi\,\mathbf{u}_B^{(\alpha)}$, where $\phi(x) = x^2$ and $\psi(x) = (\alpha+2)x + 2$ (see Table 8.1). Hence, by Lemma 7.4, the sequence

$$\left\{ \left(\widehat{B}_n^{(\alpha)}\right)^{[k]} := \frac{1}{(n+1)_k} \frac{\mathrm{d}^k \widehat{B}_{n+k}^{(\alpha)}}{\mathrm{d}x^k} \right\}_{n\geq 0}$$

is a monic OPS with respect to the functional $\mathbf{u}^{[k]} := x^{2k}\mathbf{u}_B^{(\alpha)}$ for each $k \in \mathbb{N}_0$. Moreover, by Lemma 7.1, $\mathbf{u}^{[k]}$ fulfills the functional differential equation

$$D(x^2 \mathbf{u}^{[k]}) = \psi_k\, \mathbf{u}^{[k]},$$

where $\psi_k := \psi + k\phi' = (\alpha + 2k + 2)x + 2$. Thus, $\mathbf{u}^{[k]} = \mathbf{u}_B^{(\alpha+2k)}$, $\left(\widehat{B}_n^{(\alpha)}\right)^{[k]} \equiv \widehat{B}_n^{(\alpha+2k)}$, and so

$$\frac{\mathrm{d}^k}{\mathrm{d}x^k}\left(\widehat{B}_n^{(\alpha)}(x)\right) = (n-k+1)_k\, \widehat{B}_{n-k}^{(\alpha+2k)}(x), \quad n \geq k. \tag{10.42}$$

Therefore, using McLaurin formula, we may write

$$\widehat{B}_n^{(\alpha)}(x) = \sum_{k=0}^{n} \frac{\frac{\mathrm{d}^k \widehat{B}_n^{(\alpha)}}{\mathrm{d}x^k}(0)}{k!} x^k = \sum_{k=0}^{n} \binom{n}{k} \widehat{B}_{n-k}^{(\alpha+2k)}(0)\, x^k. \tag{10.43}$$

Thus we need to compute $\widehat{B}_n^{(\alpha)}(0)$ for each $n \in \mathbb{N}_0$. Taking $z = x = 0$ in the ordinary differential equation (10.16) for $y_n := \widehat{B}_n^{(\alpha)}(x)$, and since, by (10.42) with $k = 1$,

$$y_n'(x) = n\widehat{B}_{n-1}^{(\alpha+2)}(x),$$

we obtain (for each $n \geq 1$)

$$\widehat{B}_n^{(\alpha)}(0) = \frac{2}{n+\alpha+1}\,\widehat{B}_{n-1}^{(\alpha+2)}(0),$$

hence, by iteration of this identity, we deduce

$$\widehat{B}_n^{(\alpha)}(0) = \frac{2^n}{(n+\alpha+1)_n}\,\widehat{B}_0^{(\alpha+2n)}(0),$$

i.e.,

$$\widehat{B}_n^{(\alpha)}(0) = \frac{2^n}{(n+\alpha+1)_n}. \tag{10.44}$$

Finally, inserting (10.44) into (10.43) we obtain

$$\widehat{B}_n^{(\alpha)}(x) = \frac{2^n}{(n+\alpha+1)_n}\sum_{k=0}^{n}\binom{n}{k}(n+\alpha+1)_k\left(\frac{x}{2}\right)^k. \tag{10.45}$$

A standard normalization for the Bessel polynomials is $\{Y_n^{(\alpha)}\}_{n\geq 0}$ chosen so that $Y_n^{(\alpha)}(0) = 1$, and so

$$Y_n^{(\alpha)}(x) = \sum_{k=0}^{n}\binom{n}{k}(n+\alpha+1)_k\left(\frac{x}{2}\right)^k. \tag{10.46}$$

Remark 10.3. *Although $\mathbf{u}_B^{(\alpha)}$ is not a positive definite functional, the Bessel polynomials $\widehat{B}_n^{(\alpha)}$ fulfill the following (ordinary) Rodrigues formula for $x \in \mathbb{C}\setminus\{0\}$*

$$\widehat{B}_n^{(\alpha)}(x) = \frac{1}{(n+\alpha+1)_n}\frac{1}{\omega_\alpha(x)}\frac{\mathrm{d}^n}{\mathrm{d}x^n}\big(x^{2n}\omega_\alpha(x)\big), \quad \omega_\alpha(x) := x^\alpha e^{-2/x}. \tag{10.47}$$

This formula can be used to derive the explicit expression (10.45).

Remark 10.4. *The explicit expressions* (10.29) *and* (10.35) *for Jacobi and Laguerre polynomials can be deduced using the technique we have applied to derive* (10.45). *Thus we may remove the restrictions $\alpha > -1$*

and $\beta > -1$ considered on the proof of (10.29) and (10.35), and hence these explicit formulas remain true requiring only that the corresponding functionals are regular (not necessarily positive definite).

Exercises

1. Prove relations (10.34) and (10.36).

2. Prove that the Jacobi polynomials admit the explicit representation

$$P_n^{(\alpha,\beta)}(x) = \binom{2n+\alpha+\beta}{n} \sum_{k=0}^{n} \frac{\binom{n}{k}\binom{n+\alpha}{n-k}}{\binom{2n+\alpha+\beta}{n-k}} \left(\frac{x-1}{2}\right)^k, \quad n \in \mathbb{N}_0.$$

3. (a) Suppose that $\mathbf{u} \in \mathcal{P}'$ is regular and let $\{P_n\}_{n\geq 0}$ be the corresponding monic OPS. Set $\mathbf{v} := (x-c)\mathbf{u}$, $c \in \mathbb{C}$. Prove that $\mathbf{v}$ is regular if and only if $P_n(c) \neq 0$ for all $n \in \mathbb{N}$. Under such conditions, $\{Q_n\}_{n\geq 0}$ being the monic OPS with respect to $\mathbf{v}$, show that

$$Q_n(x) = \frac{1}{x-c}\left(P_{n+1}(x) - \frac{P_{n+1}(c)}{P_n(c)} P_n(x)\right).$$

   (b) Using the results in (a), prove that the following relation among (standard) Jacobi polynomials holds:

$$(2n+\alpha+\beta+2)(x+1)P_n^{(\alpha,\beta+1)}(x)$$
$$= 2(n+\beta+1)P_n^{(\alpha,\beta)}(x) + 2(n+1)P_{n+1}^{(\alpha,\beta)}(x), \quad n \in \mathbb{N}_0.$$

4. Let $\mathbf{u}_B^{(\alpha)}$, $\alpha \in \mathbb{C} \setminus \{-2,-3,-4,\cdots\}$, be the (canonical) Bessel functional and let $\{\widehat{B}_n^{(\alpha)}\}_{n\geq 0}$be the monic OPS with respect to $\mathbf{u}_B^{(\alpha)}$. Show that the following hold:

   (a) For each $n \in \mathbb{N}_0$, $\widehat{B}_n^{(\alpha)}$ fulfills the Rodrigues formula (10.47).

   (b) If $\alpha \in \mathbb{N}_0 \cup \{-1\}$, then $\{\widehat{B}_n^{(\alpha)}\}_{n\geq 0}$ satisfies the orthogonality relations

$$\frac{1}{2\pi i}\int_{\mathbb{S}^1} \widehat{B}_m^{(\alpha)}(z)\widehat{B}_n^{(\alpha)}(z)\omega_\alpha(z)\,\mathrm{d}z$$
$$= \frac{(-1)^{n+\alpha+1}\,2^{2n+\alpha+1}\,n!}{(2n+\alpha+1)!\,(n+\alpha+1)_n}\delta_{m,n}, \quad m \in \mathbb{N}_0.$$

5. Let $\mathbf{u} \in \mathcal{P}'$ be a regular functional fulfilling the generalized functional differential equation

$$2D\big((x^2+2x+1)\mathbf{u}\big) = (-2x^2-x+1)\mathbf{u}$$

and such that $u_1 = -u_0/2$ (where, as usual, $u_n := \langle \mathbf{u}, x^n \rangle$, $n \in \mathbb{N}_0$). Show that $\mathbf{u}$ is a classical functional, identifying $\mathbf{u}$ as well as the corresponding monic OPS. Conclude that $\mathbf{u}$ is a positive definite functional uniquely represented by a positive Borel measure μ with finite moments of all orders and $\text{supp}(\mu) = [-1, +\infty)$. Determine μ explicitly.

6. (a) Prove that the integral representation (10.11) for the Jacobi functional $\mathbf{u}_J^{(\alpha,\beta)}$ is still valid provided that $\Re\alpha > -1$ and $\Re\beta > -1$.

   (b) Prove that the integral representation (10.8) for the Laguerre functional $\mathbf{u}_L^{(\alpha)}$ is still valid provided that $\Re\alpha > -1$.

   (Hint. Use the identity principle for complex analytic functions regarding α and β as complex variables.)

Chapter 11

Hypergeometric Series

In this chapter we give a short introduction to hypergeometric series and functions, preparing the reader for future studies on the topics addressed in this book. The following sections are mainly based on Chapters 1 and 2 in [2] by Andrews, Askey, and Roy (which contains much more information concerning this topic), although at times we have also followed the books [48] by Rainville, [5] by Bailey, [54] by Whittaker and Watson, [28] by Lebedev, as well as the Batman Manuscript Project [15], and the Maroni article [37]. Before introducing such a series, we need to review two other basic functions, namely the gamma and the beta functions.

11.1 The Gamma and Beta Functions

Let us start with a well-known definition.

Definition 11.1. *The gamma function is defined as*

$$\Gamma(z) := \lim_{\substack{n\to+\infty \\ (n\in\mathbb{N})}} \frac{n!\,n^{z-1}}{(z)_n}, \quad z \in \mathbb{C} \setminus \{0,-1,-2,-3,\dots\}. \tag{11.1}$$

The gamma function is a generalization of the factorial. Indeed, assuming that the above limit exists, we may write

$$\Gamma(z+1) := \lim_{n\to\infty} \frac{n!\,n^{z}}{(z+1)_n} = \lim_{n\to\infty} \frac{n!\,n^{z-1}}{(z)_n}\frac{zn}{z+n} = z\Gamma(z),$$

and so the following property holds (difference equation for the gamma function):

$$\Gamma(z+1) = z\Gamma(z), \quad z \in \mathbb{C} \setminus \{0,-1,-2,-3,\dots\}. \tag{11.2}$$

In particular, since

$$\Gamma(1) = \lim_{n\to\infty} n!\frac{n^0}{(1)_n} = \lim_{n\to\infty} \frac{n!}{n!} = 1,$$

DOI: 10.1201/9781003432067-11

we deduce

$$\Gamma(n+1) = n!, \quad n \in \mathbb{N}_0. \tag{11.3}$$

Notice also that (11.2) allows us to write the following useful identity

$$(a)_n = \frac{\Gamma(n+a)}{\Gamma(a)}, \quad a \in \mathbb{C} \setminus \{0,-1,-2,-3,\dots\}, \quad n \in \mathbb{N}_0. \tag{11.4}$$

The next theorem shows that indeed the limit defining the gamma function exists. We need to recall the definition of the Euler-Mascheroni constant:

$$\gamma := \lim_{n\to+\infty} \left(1 + \frac{1}{2} + \cdots + \frac{1}{n} - \ln n\right) = 0.5772156\dots. \tag{11.5}$$

This constant is often referred as Euler's constant. The arithmetic nature of Euler's constant is unknown. It is conjectured that "$\gamma \notin \mathbb{Q}$". In fact, it is expected that "γ is a transcendental number". This is an old and important conjecture that fits into the class of problems related with Hilbert's seventh problem. Notice that this limit exists. In fact, setting

$$u_n := \int_0^1 \frac{t}{n(n+t)}\,\mathrm{d}t = \frac{1}{n} - \ln\frac{n+1}{n},$$

we have

$$0 < u_n \le \int_0^1 \frac{1}{n(n+0)}\,\mathrm{d}t = \frac{1}{n^2}, \quad n \in \mathbb{N}.$$

Hence $\sum_{n=1}^{\infty} u_n$ converges, and so

$$1 + \frac{1}{2} + \cdots + \frac{1}{n} - \ln n = \sum_{k=1}^{n} u_k + \ln\frac{n+1}{n} \xrightarrow{n\to+\infty} \sum_{k=1}^{\infty} u_k = \gamma.$$

Theorem 11.1. *The limit (11.1) exists and it is never zero. Moreover, Γ is an analytic function in all its domain $\mathbb{C} \setminus \{0,-1,-2,-3,\dots\}$, with simple poles at the points $0,-1,-2,-3,\dots$. In addition, the identities*

$$\Gamma(z) = \frac{1}{z}\prod_{n=1}^{\infty}\left(\left(1+\frac{z}{n}\right)^{-1}\left(1+\frac{1}{n}\right)^{z}\right), \tag{11.6}$$

$$\frac{1}{\Gamma(z)} = z e^{\gamma z}\prod_{n=1}^{\infty}\left(\left(1+\frac{z}{n}\right)e^{-z/n}\right), \tag{11.7}$$

hold for each $z \in \mathbb{C}\setminus\{0,-1,-2,-3,\dots\}$, where γ is the Euler-Mascheroni constant.

Proof. We present a proof that does not assume prior knowledge of the theory of infinite products, following the exposition at the beginning of chapter XII in [54]. However, assuming some basic facts concerning this theory, a more concise proof could be written. Let $N \in \mathbb{N}$ and $z \in \mathbb{C}$ such that $|z| \leq N/2$. Recall that, taking the principal value of $\log(1+w)$, we have

$$\log(1+w) = \sum_{k=1}^{\infty} \frac{(-1)^{k-1}}{k} w^k, \quad |w| < 1.$$

Hence

$$\left|\log\left(1+\frac{z}{n}\right) - \frac{z}{n}\right| \leq \frac{|z|^2}{n^2} \sum_{k=0}^{\infty} \left|\frac{z}{n}\right|^k \leq \frac{N^2}{4n^2} \sum_{k=0}^{\infty} \left(\frac{1}{2}\right)^k = \frac{N^2}{2n^2}, \quad n > N.$$

Since the series $\displaystyle\sum_{n=N+1}^{\infty} \frac{N^2}{2n^2}$ is convergent, then Weierstrass M–test ensures that

$$\Sigma_N(z) := \sum_{n=N+1}^{\infty} \left(\log\left(1+\frac{z}{n}\right) - \frac{z}{n}\right)$$

is an absolutely and uniformly convergent series in the region $|z| \leq 1/2\,N$, and so, since its terms are analytic functions in this region, then Σ_N is an analytic function in the same region. Consequently, its exponential

$$\Lambda_N(z) := e^{\Sigma_N(z)} = \lim_{m\to\infty} \prod_{n=N+1}^{m} \left(\left(1+\frac{z}{n}\right)e^{-\frac{z}{n}}\right) =: \prod_{n=N+1}^{\infty} \left(\left(1+\frac{z}{n}\right)e^{-\frac{z}{n}}\right)$$

is an analytic function in the region $|z| \leq 1/2\,N$ which is never zero there (since it is the exponential of a finite complex value, for each N and z), and so

$$\begin{aligned}\Lambda_N(z) \prod_{n=1}^{N} \left(\left(1+\frac{z}{n}\right)e^{-\frac{z}{n}}\right) &= \lim_{m\to\infty} \prod_{n=1}^{m} \left(\left(1+\frac{z}{n}\right)e^{-\frac{z}{n}}\right) \\ &=: \prod_{n=1}^{\infty} \left(\left(1+\frac{z}{n}\right)e^{-\frac{z}{n}}\right) =: \Lambda(z)\end{aligned}$$

is an analytic function in the region $|z| \leq 1/2\,N$ which fulfills $\Lambda(z) \neq 0$ for each z in this region that does not coincide with a nonnegative integer number. Therefore, since we can take N arbitrarily large, we conclude that Λ is analytic in $\mathbb{C}$ (an entire function) and fulfills $\Lambda(z) \neq 0$ for each $z \in \mathbb{C}\backslash\{-1,-2,-3,\ldots\}$. Clearly, the zeros of $\Lambda(z)$ are precisely the numbers $-1,-2,-3,\ldots$, which are simple zeros. Now, we may write

$$e^{\gamma z}\Lambda(z) = \lim_{n\to\infty} e^{\left(1+\frac{1}{2}+\cdots+\frac{1}{n}-\ln n\right)z} \cdot \lim_{n\to\infty}\prod_{j=1}^{n}\left(\left(1+\frac{z}{j}\right)e^{-\frac{z}{j}}\right)$$

$$= \lim_{n\to\infty}\left(e^{\left(1+\frac{1}{2}+\cdots+\frac{1}{n}-\ln n\right)z} \cdot \prod_{j=1}^{n}\left(\left(1+\frac{z}{j}\right)e^{-\frac{z}{j}}\right)\right)$$

$$= \lim_{n\to\infty} n^{-z}\cdot\prod_{j=1}^{n}\left(1+\frac{z}{j}\right),$$

and, since for each $n \in \mathbb{N}$ and $z \in \mathbb{C}\setminus\{-1,-2,-3,\dots,-(n-1)\}$, we have

$$\frac{n!\,n^{z-1}}{(z)_n} = \frac{1}{z}\frac{n^z}{\frac{1+z}{1}\frac{2+z}{2}\cdots\frac{n-1+z}{n-1}} = \frac{1}{z}\left(1+\frac{z}{n}\right)\left(n^{-z}\prod_{j=1}^{n}\left(1+\frac{z}{j}\right)\right)^{-1}, \tag{11.8}$$

we conclude that there exists, and it is nonzero, the limit

$$\Gamma(z) := \lim_{n\to+\infty}\frac{n!\,n^{z-1}}{(z)_n} = \frac{1}{ze^{\gamma z}\Lambda(z)}, \quad z \in \mathbb{C}\setminus\{0,-1,-2,-3,\dots\}. \tag{11.9}$$

Moreover, taking into account the facts proved above regarding the function Λ, it follows immediately that Γ is an analytic function on $\mathbb{C}\setminus\{0,-1,-2,-3,\dots\}$ and it has simple poles at the points $0,-1,-2,-3,\dots$. Notice that (11.9) also proves (11.7). Finally, for each $z \in \mathbb{C}\setminus\{-1,-2,-3,\dots,-(n-1)\}$, (11.8) can be rewritten as

$$\frac{n!\,n^{z-1}}{(z)_n} = \left(\frac{n}{n+1}\right)^z\left(1+\frac{z}{n}\right)\frac{1}{z}\prod_{j=1}^{n}\left(\left(1+\frac{z}{j}\right)^{-1}\left(1+\frac{1}{j}\right)^z\right),$$

hence taking the limit as $n\to\infty$ we obtain (11.6). □

Historically, the gamma function was first defined by Euler, as in (11.6) above, where "$\Gamma(z)$" is the notation introduced by Legendre in 1814.

Definition 11.2. *The beta integral is*

$$B(x,y) := \int_0^1 t^{x-1}(1-t)^{y-1}\,\mathrm{d}t, \quad \Re x > 0, \quad \Re y > 0. \tag{11.10}$$

The beta function is obtained from the beta integral by analytic continuation, and we still denote it by $B(x,y)$.

Notice that the integral in (11.10) is symmetric in x and y, i.e.,

$$B(x,y) = B(y,x), \quad \Re x > 0, \quad \Re y > 0. \tag{11.11}$$

This identity follows immediately from (11.10) making the change of variables $t = 1 - s$.

Theorem 11.2. *The beta function fulfills*

$$B(x,y) = \frac{\Gamma(x)\Gamma(y)}{\Gamma(x+y)}, \quad x, y, x+y \in \mathbb{C} \setminus \{0, -1, -2, -3, \ldots\}. \tag{11.12}$$

Proof. If $\Re x > 0$ and $\Re y > 0$, the beta function $B(x,y)$ is given by (11.10). On the other hand, by Theorem 11.1, $\Gamma(x)$, $\Gamma(y)$, and $\Gamma(x+y)$ are well-defined and non-zero for all x and y such that $x, y, x+y \in \mathbb{C} \setminus \{0, -1, -2, -3, \ldots\}$. Therefore, the right-hand side of (11.12) is well-defined, and we only need to prove (11.12) for x and y such that $\Re x > 0$ and $\Re y > 0$, since—taking into account Theorem 11.1 again—the right-hand side of (11.12) provides the analytic continuation of the beta integral. Assuming $\Re x > 0$ and $\Re y > 0$, we start by proving that B fulfills the functional equation

$$B(x,y) = \frac{x+y}{y} B(x, y+1). \tag{11.13}$$

By (11.10), we may write

$$B(x, y+1) = \int_0^1 t^{x-1}(1-t)^{y-1}\,\mathrm{d}t - \int_0^1 t^x(1-t)^{y-1}\,\mathrm{d}t,$$

i.e.,

$$B(x, y+1) = B(x,y) - B(x+1, y). \tag{11.14}$$

On the other hand, integration by parts yields

$$\begin{aligned} B(x,y+1) &= \int_0^1 t^{x-1}(1-t)^y\,\mathrm{d}t \\ &= \frac{t^x}{x}(1-t)^y \bigg|_{t=0}^1 + \frac{y}{x}\int_0^1 t^x(1-t)^{y-1}\,\mathrm{d}t \\ &= \frac{y}{x} B(x+1, y). \end{aligned} \tag{11.15}$$

Inserting (11.15) in the right-hand side of (11.14) proves (11.13). Iterating (11.13), we get

$$B(x,y) = \frac{(x+y)_n}{(y)_n} B(x, y+n), \quad n \in \mathbb{N}. \tag{11.16}$$

One sees (making the change of variables $t = s/n$) that

$$B(x, y+n) = \int_0^1 t^{x-1}(1-t)^{y+n-1}\,\mathrm{d}t = \frac{1}{n^x}\int_0^n s^{x-1}\left(1-\frac{s}{n}\right)^{y+n-1}\mathrm{d}s,$$

and so (11.16) may be rewritten as

$$B(x,y) = \frac{(x+y)_n}{n!\,n^{x+y-1}}\,\frac{n!\,n^{y-1}}{(y)_n}\int_0^n s^{x-1}\left(1-\frac{s}{n}\right)^{y+n-1}\mathrm{d}s, \quad n\in\mathbb{N}. \tag{11.17}$$

Now, by definition of the gamma function, we have

$$\lim_{n\to\infty}\frac{(x+y)_n}{n!\,n^{x+y-1}} = \frac{1}{\Gamma(x+y)}, \quad \lim_{n\to\infty}\frac{n!\,n^{y-1}}{(y)_n} = \Gamma(y). \tag{11.18}$$

Moreover, using Lebesgue's dominated convergence theorem and the connections between the Lebesgue and the Riemann integrals, we deduce

$$\lim_{n\to\infty}\int_0^n s^{x-1}\left(1-\frac{s}{n}\right)^{y+n-1}\mathrm{d}s = \int_0^{+\infty} s^{x-1}e^{-s}\,\mathrm{d}s. \tag{11.19}$$

Therefore, taking the limit as $n\to+\infty$ in (11.17), from (11.18) and (11.19) we obtain

$$B(x,y) = \frac{\Gamma(y)}{\Gamma(x+y)}\int_0^{+\infty} t^{x-1}e^{-t}\,\mathrm{d}t. \tag{11.20}$$

Taking $y=1$, and since $\Gamma(1)=1$, we deduce

$$\int_0^{+\infty} t^{x-1}e^{-t}\,\mathrm{d}t = \Gamma(x+1)B(x,1) = x\,\Gamma(x)\int_0^1 t^{x-1}\,\mathrm{d}t = \Gamma(x). \tag{11.21}$$

Therefore, inserting (11.21) into (11.20) gives (11.12) for $\Re x>0$ and $\Re y>0$. Thus by the considerations at the beginning of the proof, (11.12) follows for all x and y such that $x, y, x+y \in \mathbb{C}\setminus\{0,-1,-2,-3,\dots\}$. □

Corollary 11.1. *The gamma function fulfills*

$$\Gamma(x) = \int_0^{+\infty} t^{x-1}e^{-t}\,\mathrm{d}t, \quad \Re x>0. \tag{11.22}$$

Proof. Note that (11.22) is (11.21). □

The proof of the next result is left to the reader.

Corollary 11.2 (Euler Reflection Formula). *The gamma function fulfills*

$$\Gamma(z)\Gamma(1-z) = \frac{\pi}{\sin(\pi z)}, \quad z \in \mathbb{C} \setminus \mathbb{Z}. \tag{11.23}$$

11.2 Hypergeometric Series

A hypergeometric series is a series $\sum_{n=0}^{\infty} c_n$ where

$$\frac{c_{n+1}}{c_n} \quad \text{is a rational function of } n.$$

On factorizing the polynomials in n, we may write

$$\frac{c_{n+1}}{c_n} = \frac{(n+a_1)(n+a_2)\cdots(n+a_p)}{(n+b_1)(n+b_2)\cdots(n+b_q)} \frac{x}{n+1}, \tag{11.24}$$

where x is a complex number (which appears because the polynomials may be non monic) and a_j and b_j are complex parameters such that $b_j \in \mathbb{C} \setminus \{0, -1, -2, -3, \ldots\}$. Therefore,

$$c_n = c_{n-1} \frac{(a_1+n-1)(a_2+n-1)\cdots(a_p+n-1)}{(b_1+n-1)(b_2+n-1)\cdots(b_q+n-1)} \frac{x}{n}, \quad n \in \mathbb{N},$$

and by iterating this relation we obtain

$$c_n = c_0 \frac{(a_1)_n(a_2)_n\cdots(a_p)_n}{(b_1)_n(b_2)_n\cdots(b_q)_n} \frac{x^n}{n!}. \tag{11.25}$$

Thus (up to a constant factor) a hypergeometric series is a series of the form

$${}_pF_q\left(\begin{matrix} a_1, & \cdots, & a_p \\ b_1, & \cdots, & b_q \end{matrix}; x\right) := \sum_{n=0}^{\infty} \frac{(a_1)_n(a_2)_n\cdots(a_p)_n}{(b_1)_n(b_2)_n\cdots(b_q)_n} \frac{x^n}{n!}, \quad x \in \mathbb{C}, \tag{11.26}$$

and, for all possible j,

$$a_j \in \mathbb{C}, \quad b_j \in \mathbb{C} \setminus \{0, -1, -2, -3, \ldots\}. \tag{11.27}$$

Remark 11.1. *Often, instead of the left-hand side of (11.26), the notations*

$$ {}_pF_q(a_1, \dots, a_p; b_1, \dots, b_q; x), \quad {}_pF_q(x), \quad {}_pF_q \tag{11.28}$$

are used, provided that there is no danger of misunderstanding the last two ones. Moreover, it may happen that in the numerator or in the denominator (or in both) of the fraction defining the general term of a hypergeometric series, no parameters a_j or b_j appear (this situation takes place when the number of corresponding parameters is $p = 0$ or $q = 0$, respectively). In this situation we write "—" instead of the a_j or b_j parameters, to indicate their absence. For instance,

$$ {}_0F_1\left(\begin{matrix} - \\ b_1 \end{matrix} ; x \right) = \sum_{n=0}^{\infty} \frac{x^n}{(b_1)_n \, n!}, $$

$$ {}_3F_0\left(\begin{matrix} a_1, \ a_2, \ a_3 \\ - \end{matrix} ; x \right) = \sum_{n=0}^{\infty} \frac{(a_1)_n (a_2)_n (a_3)_n}{n!} \, x^n. $$

Remark 11.2. *Notice that if a_j is zero or a negative integer number for some $j \in \{1, 2, \dots, p\}$, then the series on the right-hand side of (11.26) terminates, i.e., it reduces to a finite sum (and thus it becomes a polynomial in the variable x). In such a case, (11.26) is called a terminating hypergeometric series.*

Next we analyze the convergence of the hypergeometric series.

Theorem 11.3. *Let ${}_pF_q$ be the hypergeometric series defined by* (11.26). *Then:*

(i) *if $p \le q$, then ${}_pF_q(x)$ converges absolutely for each $x \in \mathbb{C}$;*

(ii) *if $p = q + 1$, then ${}_pF_q(x)$ converges absolutely if $|x| < 1$, and it diverges if $|x| > 1$ and the series does not terminate; and*

(iii) *if $p > q + 1$, then ${}_pF_q(x)$ diverges for each $x \in \mathbb{C} \setminus \{0\}$, provided that the series does not terminate.*

Proof. We will apply the ratio test. We may assume that the series does not terminate (otherwise it converges trivially). By (11.24), we may write

$$\left|\frac{c_{n+1}}{c_n}\right| = \frac{\left|1+\frac{a_1}{n}\right|\cdots\left|1+\frac{a_p}{n}\right| n^{p-(q+1)}}{\left|1+\frac{b_1}{n}\right|\cdots\left|1+\frac{b_q}{n}\right|\left|1+\frac{1}{n}\right|}|x|, \quad n \in \mathbb{N}_0. \tag{11.29}$$

Therefore, the following holds:

(i) Suppose $p \leq q$. Then, by (11.29),

$$\lim_{n\to+\infty}\left|\frac{c_{n+1}}{c_n}\right| = 0 < 1,$$

hence, by the ratio test, the series defining ${}_pF_q(x)$ converges absolutely for each $x \in \mathbb{C}$.

(ii) Suppose $p = q + 1$. Then, by (11.29),

$$\lim_{n\to+\infty}\left|\frac{c_{n+1}}{c_n}\right| = |x|,$$

hence the series ${}_pF_q(x)$ converges absolutely if $|x| < 1$, and it diverges if $|x| > 1$.

(iii) Suppose $p > q + 1$. By (11.29),

$$\lim_{n\to+\infty}\left|\frac{c_{n+1}}{c_n}\right| = +\infty,$$

hence the series ${}_pF_q(x)$ diverges for each $x \in \mathbb{C} \setminus \{0\}$. □

The case where $|x| = 1$ when $p = q + 1$ is of great interest. The next theorem gives the conditions for convergence in this case. Its proof requires the Gauss test. The Gauss test is a consequence of the Raabes test if $\epsilon \neq 1$, and of Bertrand's test if $\epsilon = 1$. Here we adapt the proofs of these tests to give a "direct" proof of Gauss's test, in order to maintain the proof of Theorem 11.4 more self-contained.

Lemma 11.1 (Gauss Test). *Let $\{a_n\}_{n\geq 1}$ be a sequence of positive numbers. Suppose that there exist $r > 1$, $N \in \mathbb{N}$, and a bounded sequence $\{C_n\}_{n\geq 1}$ such that*

$$\frac{a_n}{a_{n+1}} = 1 + \frac{\epsilon}{n} + \frac{C_n}{n^r}, \quad n \geq N. \tag{11.30}$$

Then the series $\sum_{n=1}^{\infty} a_n$ is convergent if $\epsilon > 1$, and it is divergent if $\epsilon \leq 1$.

Proof. Assume first $\epsilon < 1$. Since $\{C_n\}_{n\geq 1}$ is bounded and $r > 1$, there exists an integer $N_1 \geq N$ such that

$$\frac{|C_n}{n^{r-1}} \leq \frac{1-\epsilon}{2}, \quad n \geq N_1.$$

Therefore, from (11.30), we get

$$\frac{a_n}{a_{n+1}} \leq 1 + \frac{1+\epsilon}{2}\frac{1}{n} < \frac{n+1}{n}, \quad n \geq N_1.$$

Hence, $(n+1)a_{n+1} > na_n$ for each $n \geq N_1$. By a repeated application of this inequality, we deduce

$$a_n \geq \frac{c_1}{n} \quad \text{if} \quad n \geq N_1,$$

where $c_1 := N_1 a_{N_1} > 0$. Since the series $\sum_{n=1}^{\infty} 1/n$ is divergent, so is $\sum_{n=1}^{\infty} a_n$. Assume now $\epsilon = 1$. Since $r > 1$, we get

$$\frac{\ln x}{x^{r-1}} \to 0 \quad \text{as} \quad x \to +\infty.$$

Since $\{C_n\}_{n\geq 1}$ is bounded, also

$$C_n \ln n / n^{r-1} \to 0 \quad \text{as} \quad n \to +\infty.$$

Consequently, there exists an integer $N_2 > N$ such that $C_n \ln n/n^{r-1} \leq 1$ if $n \geq N_2$, and so, from (11.30),

$$\frac{a_n}{a_{n+1}} \leq 1 + \frac{1}{n} + \frac{1}{n \ln n} \leq \frac{(n+1)\ln(n+1)}{n \ln n}, \quad n \geq N_2. \tag{11.31}$$

The second inequality in (11.31) holds since it is equivalent to the inequality $1 \leq f(n)$, where

$$f(x) := (x+1)\ln\left(\frac{x+1}{x}\right);$$

and this last inequality holds since f is (strictly) decreasing on $(0, +\infty)$, and so $f(n) \geq \lim_{x\to+\infty} f(x) = 1$ for each $n > 1$. From (11.31),

$$a_n \geq \frac{c_2}{n \ln n}, \quad n \geq N_2,$$

where $c_2 := a_{N_2} N_2 \ln N_2 > 0$, and since the series $\sum_{n=2}^{\infty} \frac{1}{n \ln n}$ is divergent (use the integral test: $\int_2^{+\infty} \frac{dx}{x \ln x} = \ln \ln x|_2^{+\infty} = +\infty$), so is $\sum_{n=1}^{\infty} a_n$. Finally, assume $\epsilon > 1$. Arguing as before, there exists $N_1' \geq N$ such that

$$\frac{|C_n|}{n^{r-1}} \leq (\epsilon - 1)/2, \quad n \geq N_1'.$$

Hence, setting $q := (\epsilon + 1)/2$ and taking s such that $1 < s < q$, we deduce

$$\frac{a_n}{a_{n+1}} \geq 1 + \frac{\epsilon}{n} - \frac{\epsilon - 1}{2}\frac{1}{n} = 1 + \frac{q}{n} \geq \left(1 + \frac{1}{n}\right)^s, \quad n \geq N_3, \tag{11.32}$$

being N_3 an integer chosen so that $N_3 \geq \max\{N_1', \frac{2^{s+2}}{q-s}\}$. The last inequality in (11.32) holds by the binomial theorem, which allows us to write, for each $n \geq 2$,

$$\begin{aligned}\left(1 + \frac{1}{n}\right)^s &= 1 + \frac{s}{n} + \sum_{k=2}^{\infty} \binom{s}{k} \left(\frac{1}{n}\right)^k \\ &\leq 1 + \frac{s}{n} + \frac{4}{n^2} \sum_{k=0}^{\infty} \frac{(s)_k}{k!} \left(\frac{1}{2}\right)^k = 1 + \frac{s}{n} + \frac{2^{s+2}}{n^2},\end{aligned}$$

where we have used the inequality $\left|\binom{\alpha}{k}\right| \leq (|\alpha|)_k/k!$, valid for all $\alpha \in \mathbb{C}$ and $k \in \mathbb{N}$), and so the last inequality in (11.32) follows taking into account that

$$1 + \frac{q}{n} = 1 + \frac{s}{n} + \frac{(q-s)n}{n^2}.$$

From (11.32) we obtain $(n+1)^s a_{n+1} \leq n^s a_n$ if $n \geq N_3$, hence

$$a_n \leq \frac{c_3}{n^s}, \quad n \geq N_3,$$

where $c_3 := a_{N_3} N_3^s > 0$, and since $\sum_{n=1}^{\infty} n^{-s}$ is convergent, so is $\sum_{n=1}^{\infty} a_n$. □

Recall that, given two sequences of real or complex numbers $\{a_n\}_{n \geq 0}$ and $\{b_n\}_{n \geq 0}$, the notation "$a_n \sim b_n$ as $n \to +\infty$" means that $a_n/b_n \to 1$ as $n \to +\infty$. Before proving the theorem we also point out the following fact: the coefficient of x^n in the series ${}_{q+1}F_q(x)$ is

$$\frac{(a_1)_n (a_2)_n \cdots (a_{q+1})_n}{(b_1)_n (b_2)_n \cdots (b_q)_n \, n!} \sim \frac{\Gamma(b_1) \cdots \Gamma(b_q)}{\Gamma(a_1) \cdots \Gamma(a_{q+1})} n^{\sum a_j - \sum b_j - 1} \quad \text{as} \quad n \to +\infty. \tag{11.33}$$

Here we use the abbreviations

$$\sum a_j := \sum_{j=1}^{q+1} a_j, \quad \sum b_j := \sum_{j=1}^{q} b_j.$$

Relation (11.33) follows at once from the definition (11.1) of the gamma function, which gives

$$(z)_n \sim \frac{n! \, n^{z-1}}{\Gamma(z)}, \quad \text{as } n \to +\infty.$$

Theorem 11.4. *Let $|x| = 1$ and $p = q+1$ in the hypergeometric series defined by* (11.26). *Suppose that it is a nonterminating series.*

(i) *If $\Re(\sum a_j - \sum b_j) < 0$, then ${}_{q+1}F_q(x)$ converges absolutely.*

(ii) *If $0 \leq \Re(\sum a_j - \sum b_j) < 1$ and $x \neq 1$, then ${}_{q+1}F_q(x)$ converges conditionally.*

(iii) *If $\Re(\sum a_j - \sum b_j) \geq 1$, then ${}_{q+1}F_q(x)$ diverges.*

Proof. Since $|x| = 1$, then $x = e^{i\theta}$ for some $\theta \in \mathbb{R}$. Define

$$\alpha_n \equiv \alpha_n(\theta) := x^n = e^{in\theta}, \quad \beta_n := \frac{(a_1)_n \cdots (a_{q+1})_n}{(b_1)_n \cdots (b_q)_n \, n!}, \quad n \in \mathbb{N}_0.$$

Thus, we may write ${}_{q+1}F_q(x) = \sum_{n=0}^{\infty} f_n(x)$, where $f_n(x) := \alpha_n \beta_n$. Define also

$$\gamma := \frac{\Gamma(b_1) \cdots \Gamma(b_q)}{\Gamma(a_1) \cdots \Gamma(a_{q+1})}, \quad \epsilon := 1 - \Re\left(\sum a_j - \sum b_j\right).$$

Notice that, taking into account (11.33), we have

$$\big|f_n(x)\big| = |\beta_n| \sim \frac{|\gamma|}{n^{\epsilon}} \quad \text{as} \quad n \to +\infty. \tag{11.34}$$

The three cases (i), (ii), and (iii) in the statement of the theorem correspond, respectively, to $\epsilon > 1$, $0 < \epsilon \leq 1$, and $\epsilon \leq 0$. Therefore, we will analyze the convergence of the series ${}_{q+1}F_q(x)$ considering separately these three cases. Note that (11.34) gives us

$$\left|\frac{f_{n+1}(x)}{f_n(x)}\right| \sim \left(\frac{n}{n+1}\right)^{\epsilon} \longrightarrow 1$$

as $n \to +\infty$, and so one sees that the ratio test is inconclusive.

For $\epsilon \leq 0$, by (11.34), if $\lim_{n\to\infty} f_n(x)$ exists, it cannot be zero, hence the series

$$\sum_{n=0}^{\infty} f_n(x) \equiv {}_{q+1}F_q(x)$$

is divergent. This proves (iii). If $\epsilon > 1$, the series $\sum_{n=1}^{\infty} n^{-\epsilon}$ is convergent, and then, by (11.34), so is $\sum_{n=0}^{\infty} |f_n(x)|$, hence

$$\sum_{n=0}^{\infty} f_n(x) \equiv {}_{q+1}F_q(x)$$

is absolutely convergent. This proves (i). Now, suppose that $0 < \epsilon \leq 1$ and $x \neq 1$. Since $x = e^{i\theta}$, we may take $0 < \theta < 2\pi$. Thus, setting

$$S_n \equiv S_n(\theta) := \sum_{k=0}^{n-1} \alpha_k \beta_k,$$

we may ensure the convergence of the series ${}_{q+1}F_q(x) := \lim_{n\to\infty} S_n$ provided that we are able to show that this last limit exists. Indeed, by the summation by parts formula,

$$S_n = \sum_{k=0}^{n-1} \alpha_k \beta_k = A_n \beta_{n-1} - \sum_{k=1}^{n-1} A_k(\beta_k - \beta_{k-1}), \quad A_n := \sum_{j=0}^{n-1} \alpha_j.$$

The sequence $\{A_n\}_{n\geq 0}$ is bounded, since for each $n \geq 1$,

$$|A_n| = \left|\sum_{k=0}^{n-1} e^{ik\theta}\right| = \left|\frac{1 - e^{in\theta}}{1 - e^{i\theta}}\right| = \left|\frac{\sin(n\theta/2)}{\sin(\theta/2)}\right| \leq \frac{1}{\sin(\theta/2)}.$$

Moreover, taking into account (11.34), the sequence $\{\beta_n\}_{n\geq 0}$ converges to zero. It follows that $A_n\beta_{n-1} \to 0$ as $n \to \infty$. Thus, to conclude that $\lim_{n\to\infty} S_n$ exists, we need to show that

$$\sum_{n=1}^{+\infty} A_{n+1}(\beta_{n+1} - \beta_n)$$

is a convergent series. Indeed, we have

$$\begin{aligned}\beta_{n+1} - \beta_n &= \beta_{n+1}\left(1 - \frac{(n+b_1)\cdots(n+b_q)(n+1)}{(n+a_1)\cdots(n+a_q)(n+a_{q+1})}\right) \\ &= \beta_{n+1}\,\frac{cn^q + \pi_{q-1}(n)}{n^{q+1} + \pi_q(n)},\end{aligned} \tag{11.35}$$

where

$$c := \sum_{j=1}^{q+1} a_j - \sum_{j=1}^{q} b_j - 1, \quad \pi_{q-1} \in \mathcal{P}_{q-1}, \quad \pi_q \in \mathcal{P}_q.$$

Notice that $c \neq 0$, since $|c| \geq |\Re c| = \epsilon > 0$. Therefore, we deduce

$$|A_{n+1}(\beta_{n+1} - \beta_n)| \leq \frac{|\beta_{n+1}|}{\sin(\theta/2)}\left|\frac{cn^q + \pi_{q-1}(n)}{n^{q+1} + \pi_q(n)}\right| \sim \frac{M}{n^{1+\epsilon}}, \quad M := \frac{|\gamma\, c|}{\sin(\theta/2)} > 0.$$

Hence the (absolute) convergence of the series $\sum_{n=1}^{+\infty} A_{n+1}(\beta_{n+1} - \beta_n)$ follows from the convergence of the series

$$\sum_{n=1}^{+\infty} \frac{1}{n^{1+\epsilon}}.$$

To prove that the convergence of the series ${}_{q+1}F_q(x)$ is not absolute, we need to show that the series $\sum_{n=0}^{\infty} |\beta_n|$ is divergent. This can be done using Gauss's test (Lemma 11.1), according to which (since $\epsilon \leq 1$) we may conclude that this series diverges if we can show that there exist $N \in \mathbb{N}$ and a bounded sequence $\{C_n\}_{n\geq 1}$, such that

$$|\beta_n/\beta_{n+1}| = 1 + \epsilon/n + C_n/n^2, \quad n \geq N.$$

Indeed, taking into account (11.35) and the equality $|1-z|^2 = 1 - 2\Re z + |z|^2$, valid for any complex number z, we deduce

$$\left|\frac{\beta_n}{\beta_{n+1}}\right|^2 = \left|1 - \frac{cn^q + \pi_{q-1}(n)}{n^{q+1} + \pi_q(n)}\right|^2 = 1 + \frac{2\epsilon}{n} + \frac{B_n}{n^2},$$

where

$$B_n := \left|c - \frac{c_n}{n}\right|^2 + 2\Re c_n, \quad c_n := \frac{n(c\pi_q(n) - n\pi_{q-1}(n))}{n^{q+1} + \pi_q(n)}.$$

Note that $\{B_n\}_{n\geq 1}$ is a bounded sequence (in fact, it is convergent), so there exists $B > 0$ such that $|B_n| \leq B$ for each $n \in \mathbb{N}$. Finally, using the binomial theorem, we obtain

$$\left|\frac{\beta_n}{\beta_{n+1}}\right| = \sqrt{1 + \frac{2\epsilon}{n} + \frac{B_n}{n^2}} = 1 + \frac{\epsilon}{n} + \frac{C_n}{n^2}, \quad n \geq N,$$

where N is an integer number choosen large enough such that

$$|2\epsilon/n + B_n/n^2| \leq 1/2, \quad n \geq N,$$

and

$$C_n := \frac{B_n}{2} + \left(2\epsilon + \frac{B_n}{n}\right)^2 \sum_{k=2}^{\infty} \binom{1/2}{k} \left(\frac{2\epsilon}{n} + \frac{B_n}{n^2}\right)^{k-2}, \quad n \geq N.$$

(It does not matter how to define $C_1, \ldots, C_{N-1}$.) Clearly, $\{C_n\}_{n\geq 1}$ is bounded, since

$$|C_n| \leq \frac{B}{2} + 4(2+B)^2 \sum_{k=0}^{\infty} \frac{\left(\frac{1}{2}\right)_k}{k!} \left(\frac{1}{2}\right)^k = \frac{B}{2} + 4\sqrt{2}\,(2+B)^2, \quad n \geq N,$$

and the proof of (ii) is complete. □

Many elementary functions have representations as hypergeometric series.

We present some simple examples:

$$e^x = {}_0F_0\left(\begin{matrix}\text{—}\\ \text{—}\end{matrix};-x\right), \qquad \log(1-x) = -x\,{}_2F_1\left(\begin{matrix}1,\ 1\\ 2\end{matrix};x\right),$$

$$\sin x = x{}_0F_1\left(\begin{matrix}\text{—}\\ \frac{3}{2}\end{matrix};-x^2/4\right), \qquad \cos x = {}_0F_1\left(\begin{matrix}\text{—}\\ \frac{1}{2}\end{matrix};-x^2/4\right),$$

$$\arcsin x = x\,{}_2F_1\left(\begin{matrix}\frac{1}{2},\ \frac{1}{2}\\ \frac{3}{2}\end{matrix};x^2\right), \qquad \arctan x = x{}_2F_1\left(\begin{matrix}\frac{1}{2},\ 1\\ \frac{3}{2}\end{matrix};-x^2\right).$$

Finally, we note that the binomial theorem can be written in hypergeometric form:

$$(1-x)^{-a} = {}_1F_0\left(\begin{matrix}a\\ \text{—}\end{matrix};x\right) = \sum_{n=0}^{\infty}\frac{(a)_n}{n!}x^n, \quad |x|<1, \quad a\in\mathbb{C}. \tag{11.36}$$

11.3 The Hypergeometric Function

11.3.1 Basic Properties

The preceding example involving $\log(1-x)$ shows that although the series converges for $|x| < 1$, it has an analytic continuation as a single-valued function in the complex plane from which a line joining 1 to ∞ is deleted. We will see that this behavior describes the general situation, i.e., a ${}_2F_1$ series has a continuation to the complex plane with branch points at 1 and ∞.

Definition 11.3. *The hypergeometric function is defined by the series*

$${}_2F_1\left(\begin{matrix}a,\ b\\ c\end{matrix};x\right) := \sum_{n=0}^{\infty}\frac{(a)_n(b)_n}{(c)_n}\frac{x^n}{n!} \tag{11.37}$$

for $|x| < 1$, and, by analytic, continuation elsewhere.

Of course, in definition (11.37) it is implicitly assumed that $a, b \in \mathbb{C}$ and $c \in \mathbb{C}\setminus\{0,-1,-2,\dots\}$. Usually we reserve the use of the words hypergeometric function for ${}_2F_1$, and the hypergeometric series will be the series ${}_pF_q$ defined by (11.26)—which includes ${}_2F_1$, but will not necessarily mean just

${}_2F_1$. Notice that Theorems 11.3 and 11.4 applied to the specific ${}_2F_1$ series give the following result:

Theorem 11.5. *Consider the hypergeometric series* (11.37), *and suppose that it is a nonterminating series.*

1. *If* $|x| < 1$, *then the series is absolutely convergent.*
2. *If* $|x| > 1$, *then the series is divergent.*
3. *If* $|x| = 1$, *then the following hold:*
   (a) *If* $\Re(a+b-c) < 0$, *then the series converges absolutely;*
   (b) *if* $0 \leq \Re(a+b-c) < 1$ *and* $x \neq 1$, *then the series converges conditionally; and*
   (c) *if* $\Re(a+b-c) \geq 1$, *then the series diverges.*

Consider, for instance, $a = b = c = 1$. Then, since $(1)_n = n!$, one has

$${}_2F_1\left(\begin{matrix}1,\ 1\\ 1\end{matrix};x\right) = \sum_{n=0}^{\infty} x^n = \frac{1}{1-x} \quad \text{if } |x| < 1.$$

In this case, the series is convergent if $|x| < 1$, and it is divergent otherwise. Clearly the function $1/(x-1)$ provides the analytic continuation to $\mathbb{C} \setminus \{1\}$, and thus the hypergeometric function ${}_2F_1\left(\begin{smallmatrix}1,\ 1\\ 1\end{smallmatrix};x\right)$ becomes defined for each $x \in \mathbb{C} \setminus \{1\}$. Next, we state some important results concerning the hypergeometric function ${}_2F_1$, including the Euler integral representation and the Gauss theorem, as well as two other results involving terminating series ${}_2F_1(x)$ and ${}_3F_2(x)$ at the point $x = 1$.

11.3.2 The Euler Integral Representation

The Euler integral representation may be viewed as the analytic continuation of (11.37), provided that the condition $\Re c > \Re b > 0$ is satisfied. This condition involves only the parameters b and c, and not the parameter a, which is involved in the function $(1-xt)^{-a}$ that appears in the integrand of the integral representation—see (11.39) below. Regarded as a function of the complex variable x ($0 \leq t \leq 1$ fixed), this function is in general multivalued (it is single-valued if a is an integer number—see (11.38) below). Taking its principal value, we obtain a single-valued function which is analytic in the $x-$plane cut along the real axis from 1 to ∞, i.e., it is an analytic function of the variable x in $\mathbb{C} \setminus [1, +\infty)$. To see why this holds, we recall that, if α is a (fixed) complex number, the function defined for $z \in \mathbb{C} \setminus \{0\}$ by $z^\alpha := e^{\alpha \log_r z}$, where

$$\log_r z := \ln|z| + i \arg_r z, \quad \arg_r z \in (r, r+2\pi],$$

and $r \in \mathbb{R}$ (fixing the branch of the logarithm), is an analytic function on $\mathbb{C} \setminus \ell_r$, where ℓ_r is the ray $\ell_r := \{\rho e^{ir} \,|\, \rho \geq 0\}$. Using this fact one sees that for its principal value (which is obtained for $r = -\pi$), the function $(1-xt)^{-a}$, for fixed $t \in [0,1]$, is analytic outside the range of values $x \in \mathbb{C}$ such that the condition $1 - xt \in \ell_{-\pi} := (-\infty, 0]$ holds. This condition is impossible if $t = 0$, hence $(1-xt)^{-a}$ is analytic in $\mathbb{C}$ if $t = 0$. If $t \in (0,1]$, then $1 - xt \in (-\infty, 0]$ if and only if $x \geq 1/t$, and so $(1-xt)^{-a}$ is analytic in $\mathbb{C} \setminus [1/t, +\infty)$. The choice of the (principal) branch implies the following explicit expression of $(1-xt)^{-a}$ as single-valued function (of the variable x):

$$
\begin{aligned}
(1-xt)^{-a} &:= |1-xt|^{-\Re a} e^{\Im a \cdot \arg(1-xt) - i(\Re a \cdot \arg(1-xt) + \Im a \cdot \ln|1-xt|)}, \\
& x \in \mathbb{C} \setminus [1,+\infty), \quad t \in [0,1], \quad \arg(1-xt) \in]-\pi, \pi].
\end{aligned}
\tag{11.38}
$$

Notice that $1 - xt = 1$ for $t = 0$ and $1 - xt \neq 0$ for $0 < t \leq 1$ and each $x \in \mathbb{C} \setminus [1, +\infty)$, hence $1 - xt \neq 0$ for each $x \in \mathbb{C} \setminus [1, +\infty)$ if $t \in [0,1]$, and so $(1-xt)^{-a}$ is well defined for every $x \in \mathbb{C} \setminus [1, +\infty)$ and $t \in [0,1]$, whatever the choice of $a \in \mathbb{C}$. Note also that from this we obtain $(1-xt)^{-a} \equiv 1$ if $t = 0$ and $(1-xt)^{-a} \to 1$ as $t \to 0^+$. In conclusion: $(1-xt)^{-a}$ with its principal value defines a single-valued function analytic in the $x-$plane cut along the real axis from 1 to $+\infty$, whatever the value of $t \in [0,1]$.

Theorem 11.6 (Euler Integral Representation). *If $\Re c > \Re b > 0$, then*

$$
{}_2F_1\left(\begin{matrix} a, \; b \\ c \end{matrix}; x\right) = \frac{\Gamma(c)}{\Gamma(b)\Gamma(c-b)} \int_0^1 t^{b-1}(1-t)^{c-b-1}(1-xt)^{-a}\, dt \tag{11.39}
$$

in the $x-$plane cut along the real axis from 1 to $+\infty$. Here it should be understood that $\arg t = \arg(1-t) = 0$ and $(1-xt)^{-a}$ as its principal value.

Proof. Fix $x \in \mathbb{C}$ such that $|x| < 1$. According with the binomial theorem (11.36),

$$
t^{b-1}(1-t)^{c-b-1}(1-xt)^{-a} = \sum_{n=0}^{\infty} f_n(t), \tag{11.40}
$$

where $f_n \equiv f_n(\cdot; x) : (0,1) \to \mathbb{C}$ (regarded as a function of t) is defined by

$$
f_n(t) := \frac{(a)_n \, x^n}{n!} \, t^{n+b-1}(1-t)^{c-b-1}.
$$

Notice that $f_n \in L^1(0,1)$. Indeed, for each $t \in (0,1)$, we may write

$$
|f_n(t)| = \frac{|(a)_n| \, |x|^n}{n!} \, |t^n| \, t^{\Re b - 1}(1-t)^{\Re(c-b)-1} \leq \frac{(|a|)_n \, |x|^n}{n!} \, t^{\Re b - 1}(1-t)^{\Re(c-b)-1},
$$

the last inequality being justified by the obvious inequality $|(\alpha)_n| \leq (|\alpha|)_n$, which holds for all $\alpha \in \mathbb{C}$ and $n \in \mathbb{N}_0$. Since, by assumption, $\Re b > 0$ and $\Re(c-b) > 0$, the function $t \in (0,1) \mapsto t^{\Re b-1}(1-t)^{\Re(c-b)-1}$ is in $L^1(0,1)$. To see why this holds, notice simply that the integral of such a function is the beta integral (cf. Definition 11.2)

$$\int_0^1 t^{\Re b-1}(1-t)^{\Re(c-b)-1}\,\mathrm{d}t = B\left(\Re b, \Re(c-b)\right).$$

Moreover, we may write

$$\int_0^1 |f_n(t)|\,\mathrm{d}t \leq \frac{(|a|)_n\,|x|^n}{n!}\,B\left(\Re b, \Re(c-b)\right), \quad n \in \mathbb{N}_0.$$

Summing up for $n = 0, 1, 2, \ldots$, and noticing that, taking into account (ii) in Theorem 11.3, the series

$$\sum_{n=0}^{\infty} \frac{(|a|)_n}{n!}\,|x|^n = {}_1F_0\left(\begin{matrix}|a|\\ -\end{matrix}; |x|\right)$$

is convergent, and so

$$\sum_{n=0}^{\infty} \int_0^1 |f_n(t)|\,\mathrm{d}t \leq B\left(\Re b, \Re(c-b)\right)\,{}_1F_0\left(\begin{matrix}|a|\\ -\end{matrix}; |x|\right) < \infty. \tag{11.41}$$

Now, integrating both sides of (11.40) with respect to the variable t, (11.41) allows us to perform the change in the order of integration and summation. This yields

$$\int_0^1 t^{b-1}(1-t)^{c-b-1}(1-xt)^{-a}\,\mathrm{d}t = \sum_{n=0}^{\infty} \frac{(a)_n}{n!}x^n \int_0^1 t^{n+b-1}(1-t)^{c-b-1}\,\mathrm{d}t. \tag{11.42}$$

By (11.10) and (11.12), and taking into account (11.4), we may write

$$\begin{aligned}\int_0^1 t^{n+b-1}(1-t)^{c-b-1}\,\mathrm{d}t = B(n+b, c-b) &= \frac{\Gamma(n+b)\Gamma(c-b)}{\Gamma(n+c)}\\ &= \frac{(b)_n\Gamma(b)\Gamma(c-b)}{(c)_n\Gamma(c)}.\end{aligned}$$

Inserting this into the right-hand side of (11.42) yields (11.39) for $|x| < 1$. To prove that (11.39) holds in the cut plane $\mathbb{C} \setminus [1, +\infty)$, we will show that the integral on the right-hand side of (11.39) is an analytic function of x in the cut plane. Indeed, set

$$f(x,t) := t^{b-1}(1-t)^{c-b-1}(1-xt)^{-a}, \quad x \in \mathbb{C} \setminus [1, +\infty), \quad t \in (0,1).$$

(It does not matter how we define f for $t = 0$ or $t = 1$, provided it remains analytic in the variable x.) We have already seen that for each fixed $t \in [0, 1]$ the function $x \mapsto (1 - xt)^{-a}$ is analytic in the cut plane $\mathbb{C} \setminus [1, +\infty)$, and so the same holds for f, regarded as a function of the variable x. On the other hand, it is not difficult to see that, for each fixed $x \in \mathbb{C} \setminus [1, +\infty)$, the function $t \mapsto (1 - xt)^{-a}$ is continuous on $[0, 1]$, hence it is measurable there, and then so is f, regarded as a function of the variable t. Moreover, from (11.38), it is straightforward to show that, for each $z_0 \in \mathbb{C} \setminus [1, +\infty)$, there exists $\delta \equiv \delta(z_0) > 0$ such that

$$\left|(1 - xt)^{-a}\right| \leq C(a, z_0, \delta)e^{\pi|\Im a|}, \quad x \in \overline{B}(z_0, \delta), \quad t \in [0, 1], \tag{11.43}$$

being $\overline{B}(z_0, \delta) := \{x \in \mathbb{C} : |x - z_0| \leq \delta\} \subset \mathbb{C} \setminus [1, +\infty)$ and $C(a, z_0, \delta)$ is a constant that depends only of a, z_0, and δ, and so, we obtain

$$\sup_{|x-z_0|\leq\delta} \int_0^1 \left|f(x, t)\right| \mathrm{d}t \leq C(a, z_0, \delta)\, e^{\pi|\Im a|}\, B\left(\Re b, \Re(c - b)\right) < \infty.$$

Thus, we conclude that the right-hand side of (11.39) is an analytic function of the variable x in the cut plane, hence, since we have already proved that (11.39) holds if $|x| < 1$, it follows by analytic continuation that it holds in the cut x−plane $\mathbb{C} \setminus [1, +\infty)$ as well, and the theorem follows. □

As a first application of the Euler integral representation we derive two transformation formulas of hypergeometric functions.

Corollary 11.3. *If $|x| < 1$ and $|x/(x - 1)| < 1$, then the following transformation formula holds:*

$${}_2F_1\left(\begin{matrix} a,\ b \\ c \end{matrix}; x\right) = (1 - x)^{-a}\, {}_2F_1\left(\begin{matrix} a,\ c - b \\ c \end{matrix}; \frac{x}{x - 1}\right), \tag{11.44}$$

and if $|x| < 1$, then

$${}_2F_1\left(\begin{matrix} a,\ b \\ c \end{matrix}; x\right) = (1 - x)^{c-a-b}\, {}_2F_1\left(\begin{matrix} c - a,\ c - b \\ c \end{matrix}; x\right). \tag{11.45}$$

Here, it should be understood that $(1 - x)^{-a}$ and $(1 - x)^{c-a-b}$ have their principal values. Moreover, these formulas are valid for all complex parameters a, b, and c, provided that c is neither zero nor a negative integer number.

Proof. Assume first $\Re c > \Re b > 0$. To prove (11.44), make the substitution $t = 1 - s$ in the Euler integral (11.39). Then

$$
\begin{aligned}
{}_2F_1\left(\begin{matrix} a,\ b \\ c \end{matrix}; x\right) &= \frac{\Gamma(c)}{\Gamma(b)\Gamma(c-b)} \int_0^1 (1-s)^{b-1} s^{c-b-1} (1-x+xs)^{-a}\, \mathrm{d}s \\
&= \frac{(1-x)^{-a}\Gamma(c)}{\Gamma(b)\Gamma(c-b)} \int_0^1 (1-s)^{b-1} s^{c-b-1} \left(1 - \frac{xs}{x-1}\right)^{-a} \mathrm{d}s \\
&= (1-x)^{-a}\, {}_2F_1\left(\begin{matrix} a,\ c-b \\ c \end{matrix}; \frac{x}{x-1}\right).
\end{aligned}
$$

To prove (11.45), we consider (11.44) and note that the hypergeometric series is symmetric in the parameters appearing in the numerator. Therefore, we may write

$$
{}_2F_1\left(\begin{matrix} a,\ b \\ c \end{matrix}; x\right) = (1-x)^{-a}\, {}_2F_1\left(\begin{matrix} c-b,\ a \\ c \end{matrix}; \frac{x}{x-1}\right).
$$

Applying again (11.44) (to the last ${}_2F_1$), we obtain

$$
\begin{aligned}
{}_2F_1\left(\begin{matrix} a,\ b \\ c \end{matrix}; x\right) &= (1-x)^{-a} \left(1 - \frac{x}{x-1}\right)^{-c+b} {}_2F_1\left(\begin{matrix} c-b,\ c-a \\ c \end{matrix}; \frac{\frac{x}{x-1}}{\frac{x}{x-1} - 1}\right) \\
&= (1-x)^{-a}(1-x)^{c-b}\, {}_2F_1\left(\begin{matrix} c-a,\ c-b \\ c \end{matrix}; x\right).
\end{aligned}
$$

Therefore, (11.44) and (11.45) hold under the assumption $\Re c > \Re b > 0$. Analytic continuation in the parameters b and c gives (11.44) and (11.45) for all complex values of a, b, and c, with $c \neq 0, -1, -2, -3, \ldots$, and the result follows. □

Remark 11.3. *The hypergeometric ${}_2F_1$ series defined on the right-hand side of (11.44) converges for $|x/(x-1)| < 1$. Thus, since this condition is equivalent to $\Re x < \frac{1}{2}$, the right-hand side of (11.44) gives the analytic continuation to the region $\Re x < \frac{1}{2}$ (via the Euler integral representation) of the series defined by ${}_2F_1\left(\begin{smallmatrix} a,\ b \\ c \end{smallmatrix}; x\right)$.*

11.3.3 The Gauss Summation Formula

Our next result is a celebrated theorem by Gauss. It is convenient to state firstly the following result:

Lemma 11.2. *If* $\Re(c-a-b)>0$, *then*

$$ {}_2F_1\left(\begin{matrix} a,\ b \\ c \end{matrix};1\right) = \frac{(c-a)(c-b)}{c(c-a-b)}\, {}_2F_1\left(\begin{matrix} a,\ b \\ c+1 \end{matrix};1\right). \tag{11.46} $$

Proof. Set

$$ A_n := \frac{(a)_n(b)_n}{n!(c)_n}, \quad B_n := \frac{(a)_n(b)_n}{n!(c+1)_n}, \quad n \in \mathbb{N}_0. $$

After straightforward computations, we deduce

$$ c(c-a-b)A_n = (c-a)(c-b)B_n + cnA_n - c(n+1)A_{n+1}, \quad n \in \mathbb{N}_0. $$

Therefore, summing up from $n=0$ to $n=N$, we obtain

$$ c(c-a-b)\sum_{n=0}^{N} A_n = (c-a)(c-b)\sum_{n=0}^{N} B_n - c(N+1)A_{N+1}, \quad N \in \mathbb{N}. \tag{11.47} $$

Now, as $N \to +\infty$,

$$ \sum_{n=0}^{N} A_n \to {}_2F_1\left(\begin{matrix} a,\ b \\ c \end{matrix};1\right), \quad \sum_{n=0}^{N} B_n \to {}_2F_1\left(\begin{matrix} a,\ b \\ c+1 \end{matrix};1\right). $$

Moreover, by (11.33), as $N \to +\infty$,

$$ \frac{(a)_{N+1}(b)_{N+1}}{(c)_{N+1}\,(N+1)!} \sim \frac{\Gamma(c)}{\Gamma(a)\Gamma(b)}\,(N+1)^{a+b-c-1}, $$

and so, since $\Re(a+b-c)<0$, we obtain

$$ (N+1)A_{N+1} = (N+1)\,\frac{(a)_{N+1}(b)_{N+1}}{(c)_{N+1}\,(N+1)!} \sim \frac{\Gamma(c)}{\Gamma(a)\Gamma(b)}\,(N+1)^{a+b-c} \to 0. $$

Therefore, taking $N \to +\infty$ in (11.47) yields (11.46). □

Theorem 11.7 (Gauss Summation Formula). *If* $\Re(c-a-b)>0$, *then*

$$ {}_2F_1\left(\begin{matrix} a,\ b \\ c \end{matrix};1\right) = \frac{\Gamma(c)\Gamma(c-a-b)}{\Gamma(c-a)\Gamma(c-b)}. \tag{11.48} $$

Proof. Iterating (11.46) n times yields

$$ {}_2F_1\left(\begin{matrix} a,\ b \\ c \end{matrix};1\right) = \frac{(c-a)_n(c-b)_n}{(c)_n(c-a-b)_n}\,{}_2F_1\left(\begin{matrix} a,\ b \\ c+n \end{matrix};1\right), \quad n\in\mathbb{N}. \tag{11.49} $$

By (11.33), as $n\to+\infty$,

$$ \frac{(c-a)_n(c-b)_n}{(c)_n(c-a-b)_n} = \frac{(c-a)_n(c-b)_n(1)_n}{(c)_n(c-a-b)_n\,n!} \sim \frac{\Gamma(c)\Gamma(c-a-b)}{\Gamma(c-a)\Gamma(c-b)\Gamma(1)}, $$

i.e., recalling that $\Gamma(1)=1$,

$$ \lim_{n\to+\infty}\frac{(c-a)_n(c-b)_n}{(c)_n(c-a-b)_n} = \frac{\Gamma(c)\Gamma(c-a-b)}{\Gamma(c-a)\Gamma(c-b)}. $$

Therefore, (11.48) will be proved taking the limit in (11.49) as $n\to\infty$, provided we are able to show that

$$ \lim_{n\to+\infty}\,{}_2F_1\left(\begin{matrix} a,\ b \\ c+n \end{matrix};1\right) = 1. \tag{11.50} $$

Let $u_k(a,b,c)$ denote the coefficient of x^k in ${}_2F_1\left(\begin{smallmatrix} a,\ b \\ c \end{smallmatrix};x\right)$, i.e., write

$$ {}_2F_1\left(\begin{matrix} a,\ b \\ c \end{matrix};x\right) = \sum_{k=0}^{\infty} u_k(a,b,c)x^k, \quad u_k(a,b,c) := \frac{(a)_k(b)_k}{(c)_k\,k!}. $$

For each $k\in\mathbb{N}_0$ and $n\in\mathbb{N}$ such that $n>|c|$, we have $|(a)_k|\leq(|a|)_k$, $|(b)_k|\leq(|b|)_k$, and $|(c+n)_k|\geq(n-|c|)_k$, hence

$$ \big|u_k(a,b,c+n)\big| = \left|\frac{(a)_k(b)_k}{(c+n)_k\,k!}\right| \leq \frac{(|a|)_k(|b|)_k}{(n-|c|)_k\,k!} = u_k\left(|a|,|b|,n-|c|\right), $$

and so we may write

$$ \left|{}_2F_1\left(\begin{matrix} a,\ b \\ c+n \end{matrix};1\right)-1\right| = \left|\sum_{k=1}^{\infty} u_k(a,b,c+n)\right| \leq \sum_{k=1}^{\infty} u_k\left(|a|,|b|,n-|c|\right). $$

Thus

$$ \left|{}_2F_1\left(\begin{matrix} a,\ b \\ c+n \end{matrix};1\right)-1\right| \leq \sum_{k=0}^{\infty} u_{k+1}\left(|a|,|b|,n-|c|\right), \quad n>|c|. \tag{11.51} $$

Next, notice that, for each $k\in\mathbb{N}_0$ and $n>|c|$,

$$ \begin{aligned} u_{k+1}\left(|a|,|b|,n-|c|\right) &= \frac{(|a|)_{k+1}(|b|)_{k+1}}{(n-|c|)_{k+1}\,(k+1)!} \\ &= \frac{1}{k+1}\,\frac{|ab|}{n-|c|}\,\frac{(|a|+1)_k(|b|+1)_k}{(n+1-|c|)_k\,k!} \\ &\leq \frac{|ab|}{n-|c|}u_k\left(|a|+1,|b|+1,n+1-|c|\right). \end{aligned} $$

Therefore, from (11.51) we obtain

$$\left|{}_2F_1\left(\begin{matrix}a,\ b\\ c+n\end{matrix};1\right)-1\right|\leq\frac{|ab|}{n-|c|}\,{}_2F_1\left(\begin{matrix}|a|+1,\ |b|+1\\ n+1-|c|\end{matrix};1\right),\quad n>|c|.$$

According to (i) in Theorem 11.4, the series ${}_2F_1\left(\begin{smallmatrix}|a|+1,\ |b|+1\\ n+1-|c|\end{smallmatrix};1\right)$ converges (absolutely) if $n>|a|+|b|+|c|+1$. This series is, clearly, a decreasing function of n, hence it is bounded by a positive number independent of n, say, $M\equiv M(a,b,c)>0$, and so

$$\left|{}_2F_1\left(\begin{matrix}a,\ b\\ c+n\end{matrix};1\right)-1\right|\leq\frac{|ab|M}{n-|c|},\quad n>|a|+|b|+|c|+1.$$

Therefore, taking the limit as $n\to+\infty$ we obtain (11.50). □

Corollary 11.4 (Chu-Vandermonde Identity).

$$_2F_1\left(\begin{matrix}-n,\ a\\ c\end{matrix};1\right)=\frac{(c-a)_n}{(c)_n},\quad n\in\mathbb{N}_0. \tag{11.52}$$

Proof. The result follows immediately taking $b=-n$ in (11.48) and using (11.4). □

11.4 The Pfaff-Saalschütz, Dixon, and Dougall Identities

The Chu-Vandermonde identity (11.52) gives a closed formula for a terminating ${}_2F_1$ hypergeometric series. Similarly, the Pfaff-Saalschütz identity gives a closed formula for a terminating ${}_3F_2$ hypergeometric series. These kinds of formulas are very useful in the computation of binomial sums in closed form, as we will see in the next section.

Theorem 11.8 (Pfaff-Saalschütz Identity). *For each* $n\in\mathbb{N}$,

$$_3F_2\left(\begin{matrix}-n,\ a,\ b\\ c,\ 1+a+b-c-n\end{matrix};1\right)=\frac{(c-a)_n(c-b)_n}{(c)_n(c-a-b)_n}. \tag{11.53}$$

Proof. By (11.45) and the binomial theorem (11.36),

$$ {}_1F_0\left({c-a-b \atop —};x\right)\cdot {}_2F_1\left({a,\ b \atop c};x\right) = {}_2F_1\left({c-a,\ c-b \atop c};x\right), \quad |x|<1. $$

Rewrite this equation as

$$ \left(\sum_{n=0}^{\infty}\frac{(c-a-b)_n}{n!}x^n\right)\left(\sum_{n=0}^{\infty}\frac{(a)_n(b)_n}{(c)_n\, n!}x^n\right) $$

$$ =\sum_{n=0}^{\infty}\frac{(c-a)_n(c-b)_n}{(c)_n\, n!}x^n, \quad |x|<1. $$

Form the Cauchy product of the series on the left-hand side and then equate the coefficients of x^n in both sides of the resulting equality. This yields

$$ \sum_{j=0}^{n}\frac{(a)_j(b)_j(c-a-b)_{n-j}}{j!\,(c)_j(n-j)_j}=\frac{(c-a)_n(c-b)_n}{(c)_n\, n!}, \quad n\in\mathbb{N}_0. \tag{11.54} $$

Now, taking into account the equalities

$$ (\alpha)_{n-j}=\frac{(-1)^j(\alpha)_n}{(1-\alpha-n)_j}, \quad \frac{(-1)^j n!}{(n-j)!}=(-n)_j, \quad j=0,\dots,n, $$

the sum on the left-hand side of (11.54) becomes

$$ \sum_{j=0}^{n}\frac{(a)_j(b)_j(-n)_j(c-a-b)_n}{j!(c)_j(1+a+b-c-n)_j\, n!} $$

$$ =\frac{(c-a-b)_n}{n!}\,{}_3F_2\left({-n,\ a,\ b \atop c,\ 1+a+b-c-n};1\right), $$

and the theorem is proved. □

In the next section we will present examples illustrating how the Chu-Vandermonde and Pfaff-Saalschütz identities can be useful to obtain closed formulas for sums involving binomial coefficients. In the applications for such binomial identities, often the case $p=q+1$ occurs, and the success of the procedure depends upon certain relations fulfilled by the parameters $a_1,\dots,a_{q+1}$ and $b_1,\dots,b_q$ appearing in the definition of

$$ {}_{q+1}F_q\left({a_1,\ \cdots,\ a_{q+1} \atop b_1,\ \cdots,\ b_q};x\right). \tag{11.55} $$

The series (11.55) is called $k-$balanced at $x=1$ if one of the a_j's is a negative integer number, and the following condition holds:

$$ k+\sum_{j=1}^{q+1}a_j=\sum_{j=1}^{q}b_j. \tag{11.56} $$

The condition that one of the a_j's is a negative integer number means that the series terminates. This condition seems artificial, but without it many results do not hold. An 1−balanced series is also called Saalschützian. The series (11.55) is called well-poised if

$$1 + a_1 = b_1 + a_2 = \cdots = b_q + a_{q+1}. \tag{11.57}$$

We conclude by stating, without proof (the proofs can be found in several of the textbooks presented in the bibliography), two theorems involving two identities of these types. The Dixon identity applies to a well-poised ${}_3F_2$ series, while the Dougall identity applies to a well-poised 2−balanced ${}_7F_6$ series.

Theorem 11.9 (Dixon Identity). *The identity*

$$
\begin{aligned}
&{}_3F_2\left(\begin{matrix} a,\ -b,\ -c \\ 1+a+b,\ 1+a+c \end{matrix};1\right) \\
&= \frac{\Gamma\left(1+\frac{a}{2}\right)\Gamma(1+a+b)\Gamma(1+a+c)\Gamma\left(1+\frac{a}{2}+b+c\right)}{\Gamma(1+a)\Gamma\left(1+\frac{a}{2}+b\right)\Gamma\left(1+\frac{a}{2}+c\right)\Gamma(1+a+b+c)}
\end{aligned}
\tag{11.58}
$$

holds, where the condition $\Re(a+2b+2c+2) > 0$ *is assumed whenever the left-hand side is an infinite series.*

Theorem 11.10 (Dougall Identity). *For each* $n \in \mathbb{N}$,

$$
\begin{aligned}
&{}_7F_6\left(\begin{matrix} -n,\ a,\ 1+a/2,\ -b,\ -c,\ -d,\ -e \\ a/2,\ 1+a+b,\ 1+a+c,\ 1+a+d,\ 1+d+e,\ 1+a+n \end{matrix};1\right) \\
&= \frac{(1+a)_n(1+a+b+c)_n(1+a+b+d)_n(1+a+c+d)_n}{(1+a+b)_n(1+a+c)_n(1+a+d)_n(1+a+b+c+d)_n},
\end{aligned}
\tag{11.59}
$$

provided that $1 + 2a + b + c + d + e + n = 0$.

11.5 Binomial Sums

One area where hypergeometric identities are very useful is in the evaluation of sums of products of binomial coefficients. The main idea behind this procedure is writing such a sum as a hypergeometric series. In this section we present three examples illustrating the power of this technique. While working on examples of this type, we need to compute quotients involving binomial coefficients, so often it is useful to make use of the following identities (easy to check), which hold for $\alpha \in \mathbb{C}$ and $k \in \mathbb{N}_0$:

$$\frac{\binom{\alpha+1}{k+1}}{\binom{\alpha}{k}} = \frac{\alpha+1}{k+1}, \qquad \frac{\binom{\alpha+2}{k+1}}{\binom{\alpha}{k}} = \frac{(\alpha+2)(\alpha+1)}{(k+1)(\alpha-k+1)},$$

$$\frac{\binom{\alpha}{k+1}}{\binom{\alpha}{k}} = \frac{\alpha-k}{k+1}, \qquad \frac{\binom{\alpha+1}{k+2}}{\binom{\alpha}{k}} = \frac{(\alpha+1)(\alpha-k)}{(k+2)(k+1)}.$$

It is also useful to keep in mind the relations

$$\binom{\alpha}{n} := \frac{\alpha(\alpha-1)\cdots(\alpha-n+1)}{n!}$$

$$= \frac{(-1)^n(-\alpha)_n}{n!} = \frac{(\alpha-n+1)_n}{n!}, \quad \alpha \in \mathbb{C}, \quad n \in \mathbb{N}.$$

As a first example, we show the following result:

Proposition 11.1.

$$\sum_{j=0}^{n} (-1)^j \frac{\binom{\alpha}{j}\binom{\alpha-1-j}{n-j}}{j+1} = \frac{\binom{\alpha}{n+1} + (-1)^n}{\alpha+1}, \quad \alpha \in \mathbb{C}\backslash\{-1\}, \quad n \in \mathbb{N}_0. \tag{11.60}$$

Proof. Denote the sum of the left-hand side by S, so that

$$S := \sum_{j=0}^{n} c_j, \quad c_j := (-1)^j \frac{\binom{\alpha}{j}\binom{\alpha-1-j}{n-j}}{j+1}.$$

To write this sum as a hypergeometric series, we first compute the ratio c_{j+1}/c_j, and then we put it in the form (11.24):

$$\frac{c_{j+1}}{c_j} = -\frac{(j+1)\binom{\alpha}{j+1}\binom{\alpha-2-j}{n-1-j}}{(j+2)\binom{\alpha}{j}\binom{\alpha-1-j}{n-j}} = \frac{(j-n)(j-\alpha)(j+1)}{(j-\alpha+1)(j+2)} \frac{1}{j+1},$$

the last equality following immediately. Thus—cf. (11.25)—,

$$S := \sum_{j=0}^{n} c_j = c_0 \sum_{j=0}^{n} \frac{(-n)_j(-\alpha)_j(1)_j}{(-\alpha+1)_j(2)_j j!}, \quad c_0 := \binom{\alpha-1}{n}, \tag{11.61}$$

and so the given binomial sum can be written in hypergeometric form as

$$S = \binom{\alpha-1}{n} {}_3F_2\left(\begin{matrix}-n,\ -\alpha,\ 1\\ -\alpha+1,\ 2\end{matrix};1\right). \tag{11.62}$$

At this point, one could try to apply the Pfaff-Saalschütz identity (11.53). However, the ${}_3F_2$ series in (11.62) is not in the form of the ${}_3F_2$ appearing in (11.53). In fact, if $a = -\alpha$, $b = 1$, and $c = -\alpha + 1$, then $1 + a + b - c - n = 1 - n \neq 2$. Thus, the Pfaff-Saalschütz identity does not apply. Nevertheless, returning to (11.61), and noting that $(1)_j = j!$, $(2)_j = (1)_{j+1} = (j+1)!$, and $(z)_j = (z-1)_{j+1}/(z-1)$, we may write

$$\begin{aligned} S &= -\binom{\alpha-1}{n} \frac{\alpha}{(n+1)(\alpha+1)} \sum_{j=0}^{n} \frac{(-n-1)_{j+1}(-\alpha-1)_{j+1}}{(-\alpha)_{j+1}(j+1)!} \\ &= -\binom{\alpha-1}{n} \frac{\alpha}{(n+1)(\alpha+1)} \left(\sum_{j=0}^{n+1} \frac{(-n-1)_j(-\alpha-1)_j}{(-\alpha)_j\, j!} - 1 \right) \\ &= -\binom{\alpha-1}{n} \frac{\alpha}{(n+1)(\alpha+1)} \left({}_2F_1\left(\begin{matrix}-(n+1),\ -\alpha-1\\ -\alpha\end{matrix};1\right) - 1 \right). \end{aligned} \tag{11.63}$$

The last ${}_2F_1$ may be computed by the Chu-Vandermonde identity (11.52), and so

$${}_2F_1\left(\begin{matrix}-(n+1),\ -\alpha-1\\ -\alpha\end{matrix};1\right) = \frac{(1)_{n+1}}{(-\alpha)_{n+1}} = \frac{(n+1)!}{(-\alpha)_{n+1}}.$$

Inserting this expression into (11.63) and simplifying the result, we obtain (11.60). □

As a second example, let us show the following result:

Proposition 11.2.

$$\sum_{j=0}^{n-m} \frac{(-1)^j}{j+1} \binom{n+j}{m+2j} \binom{2j}{j} = \binom{n-1}{m-1}, \quad m, n \in \mathbb{N}. \tag{11.64}$$

Proof. It is clear that (11.64) holds if $n < m$, since in such case both sides of (11.64) are equal to zero (this holds because $\binom{k}{\ell} = 0$ if $k, \ell \in \mathbb{N}$ with $k < \ell$). Henceforth, we assume $1 \leq m \leq n$. Let

$$S := \sum_{j=0}^{\infty} c_j, \quad c_j := \frac{(-1)^j}{j+1} \binom{n+j}{m+2j} \binom{2j}{j}.$$

Since $\binom{n+j}{m+2j} = 0$ for $n + j < m + 2j$ (i.e., if $j > n - m$), we have $c_j = 0$ for $j > n - m$, and so S is indeed a finite sum. It is easy to check that

$$\frac{c_{j+1}}{c_j} = -\frac{(j+1)\binom{n+j+1}{m+2j+2}\binom{2j+2}{j+1}}{(j+2)\binom{n+j}{m+2j}\binom{2j}{j}}$$

$$= \frac{(j+n+1)(j-n+m)\left(j+\frac{1}{2}\right)}{\left(j+\frac{m}{2}+1\right)\left(j+\frac{m+1}{2}\right)(j+2)}.$$

Thus, since $c_0 = \binom{n}{m}$, we obtain, see (11.25),

$$S = \binom{n}{m} \sum_{j=0}^{\infty} \frac{(n+1)_j (m-n)_j \left(\frac{1}{2}\right)_j}{\left(\frac{m}{2}+1\right)_j \left(\frac{m+1}{2}\right)_j (j+1)!} \tag{11.65}$$

$$= \binom{n}{m} \sum_{j=1}^{\infty} \frac{(n+1)_{j-1} (m-n)_{j-1} \left(\frac{1}{2}\right)_{j-1}}{\left(\frac{m}{2}+1\right)_{j-1} \left(\frac{m+1}{2}\right)_{j-1} j!}. \tag{11.66}$$

If $m > 1$, using $(z+1)_{j-1} = (z)_j / z$, the last sum can be written as

$$S = -\frac{1}{2}\binom{n}{m}\frac{m(m-1)}{n(m-n-1)}\left({}_3F_2\left(\begin{matrix} n,\ m-n-1,\ -\dfrac{1}{2} \\ \dfrac{m}{2},\ \dfrac{m-1}{2} \end{matrix};1\right) - 1\right). \tag{11.67}$$

Since $m < n+1$, by the Pfaff-Saalschütz identity (11.53) the last ${}_3F_2$ series becomes

$${}_3F_2\left(\begin{matrix} -(n+1-m),\ n,\ -\frac{1}{2} \\ \dfrac{m-1}{2},\ \dfrac{m}{2} \end{matrix};1\right) = \frac{\left(\dfrac{m-1}{2}-n\right)_{n+1-m}\left(\dfrac{m}{2}\right)_{n+1-m}}{\left(\dfrac{m-1}{2}\right)_{n+1-m}\left(\dfrac{m}{2}-n\right)_{n+1-m}}.$$

Inserting this expression into the right-hand side of (11.67) and simplifying the resulting equality—in this simplification process the relation

$$(\alpha - n)_k = (-1)^k(-\alpha + n - k + 1)_k$$

may be useful—, we obtain (11.64) whenever $m > 1$. If $m = 1$, noting that $(1)_j = j!$ and $(j+1)! = (2)_j$, the first equality in (11.65) gives

$$S = \binom{n}{1} {}_3F_2\left(\begin{matrix} -(n-1),\ n+1,\ \frac{1}{2} \\ \frac{3}{2},\ 2 \end{matrix};1\right) = n\,\frac{\left(\frac{1}{2}-n\right)_{n-1}(1)_{n-1}}{\left(\frac{3}{2}\right)_{n-1}(-n)_{n-1}} = 1,$$

where in the second equality we have used again the Pfaff-Saalschütz identity (11.53). This proves (11.64) for $m = 1$. Notice that since $c_j = 0$ if $j > n - m$ then in (11.64) one may replace $\sum_{j=0}^{n-m}$ by $\sum_{j=0}^{\infty}$. □

As a last example, we show the following result:

Proposition 11.3.

$$\sum_{k=1}^{\ell} 2k\binom{2p}{k+p}\binom{2n}{k+n}$$

$$= \frac{4np}{n+p}\binom{2p-1}{p}\binom{2n-1}{n}, \quad \ell := \min\{n,p\}, \quad n,p \in \mathbb{N}. \quad (11.68)$$

Proof. Denote the sum of the left-hand side of (11.68) by S. Then

$$S = \sum_{k\geq 0} c_k, \quad c_k := 2(k+1)\binom{2p}{k+1+p}\binom{2n}{k+1+n}.$$

The last equality holds, indeed, since $c_k = 0$ if $k \geq \min\{n,p\} =: \ell$. To write this sum as a (terminating) hypergeometric series, we compute the ratio c_{k+1}/c_k:

$$\frac{c_{k+1}}{c_k} = \frac{(k+2)(k+1-p)(k+1-n)}{(k+2+p)(k+2+n)}\,\frac{1}{k+1}, \quad k = 0,1,\ldots,\ell-1.$$

Thus—cf. (11.25)—, the given sum can be written in hypergeometric form as

$$S = c_0 \sum_{k\geq 0} \frac{(2)_k(1-p)_k(1-n)_k}{(2+p)_k(2+n)_k}$$

$$= 2\binom{2p}{1+p}\binom{2n}{1+n} {}_3F_2\left(\begin{matrix} 2,\ 1-p,\ 1-n \\ 2+p,\ 2+n \end{matrix};1\right).$$

This ${}_3F_2$ series can be computed using Dixon's identity (11.58), taking therein $a = 2$, $b = p - 1$, and $c = n - 1$, and so

$$S = 2\binom{2p}{1+p}\binom{2n}{1+n}\frac{\Gamma(2)\Gamma(2+p)\Gamma(2+n)\Gamma(p+n)}{\Gamma(3)\Gamma(1+p)\Gamma(1+n)\Gamma(p+n+1)}$$

$$= \binom{2p}{1+p}\binom{2n}{1+n}\frac{(1+p)(1+n)}{p+n},$$

where the last equality follows from (11.2), and so (11.68) follows. □

The reader is invited to read the very interesting article [51], where Roy presented (the above and) several other examples, pointing out the power of this technique to compute intricate binomial sums. A powerful technique to prove identities between hypergeometric functions was developed by Zeilberger and Wilf, called the creative telescoping method. This method is also referred to as the W–Z method, and it is described e.g. in the books [47] and [2]. As a matter of fact, many further developments of these techniques have appeared since then, including extensions to the so-called (basic) q-hypergeometric series as well as full algorithm implementations on the computer.

Exercises

1. Prove the limit relation (11.19).

   (Hint. Use the Lebesgue convergence-dominated theorem and the connections between the Lebesgue integral and the proper and improper Riemann integrals. It may be useful to notice that for real t and $a > 0$,

$$\left(1 - \frac{t}{n}\right)^{a+n-1} \chi_{[0,n]}(t) \leq e^{1-t}, \quad n \in \mathbb{N}.)$$

2. Prove the Euler reflection formula (11.23).

   (Hint. Set $t = s/(1+s)$ in the definition of the beta integral to obtain

$$\int_0^\infty \frac{s^{x-1}}{(1+s)^{x+y}}\,\mathrm{d}s = \frac{\Gamma(x)\Gamma(y)}{\Gamma(x+y)}, \quad \Re x > 0, \quad \Re y > 0.$$

   This gives

$$\Gamma(x)\Gamma(1-x) = \int_0^\infty \frac{t^{x-1}}{1+t}\,\mathrm{d}t, \quad 0 < x < 1.$$

   The last integral can be computed by the residue theorem using the contour integral

$$\int_{\Gamma_{\epsilon,R}} \frac{z^{x-1}}{1-z}\,\mathrm{d}z,$$

where $\Gamma_{\epsilon,R} := C_{\epsilon,R} \cup \ell^+_{\epsilon,R} \cup C_\epsilon \cup \ell^-_{\epsilon,R}$ is a closed path, $C_{\epsilon,R}$ is an incomplete circle around the origin of radius R with starting and ending points at $z = -R\cos\theta \pm i\epsilon$, not containing the point $z = -R$, being $0 < \epsilon < 1 < R$ and $\theta := \arcsin(\epsilon/R)$, C_ϵ is the semicircle around the origin of radius ϵ joining the points $z = \pm i\epsilon$ and containing $z = \epsilon$, and $\ell^\pm_{\epsilon,R}$ are two segments parallel to the negative real axis, one of them starting at $z = -R\cos\theta + i\epsilon$ and ending at $z = i\epsilon$, and the other one starting at $z = -i\epsilon$ and ending at $z = -R\cos\theta - i\epsilon$.)

3. Prove the Legendre duplication formula:

$$\sqrt{\pi}\,\Gamma(2x) = 2^{2x-1}\,\Gamma(x)\Gamma\left(x+\tfrac{1}{2}\right), \quad x \in \mathbb{C} \setminus \left\{-\frac{k}{2} : k \in \mathbb{N}_0\right\}.$$

4. Show that the estimate (11.43) holds.

   (Hint. For each $z_0 \in \mathbb{C} \setminus [1,+\infty)$, (11.43) holds if we define δ and $C(a,z_0,\delta)$ as

$$\begin{aligned}\delta &:= \frac{1}{2}\,|\Im z_0|\chi_{\mathbb{R}\setminus\{0\}}(\Im z_0)\\ &\quad + \left(\frac{1}{2}(1-z_0)\chi_{(-\infty,0]}(\Re a) + \frac{1}{6}(2 - z_0 - |z_0|)\chi_{(0,+\infty)}(\Re a)\right)\chi_{\{0\}}(\Im z_0),\end{aligned}$$

$$\begin{aligned}C(a,z_0,\delta) &:= (1+|z_0|+\delta)^{-\Re a}\,\chi_{(-\infty,0]}(\Re a)\\ &\quad + \left(1/\delta^{\Re a}\chi_{\{0\}}(\Im z_0) + (1+|z_0|/\delta)^{\Re a}\chi_{\mathbb{R}\setminus\{0\}}(\Im z_0)\right)\chi_{(0,+\infty)}(\Re a).\end{aligned}$$

   In case $\Re a > 0$, it may be useful to notice that

$$|1-xt|^2 = \begin{cases} (1-z_0t)((1-z_0t)+2t(z_0-\Re x)) + |x-z_0|^2t^2, & \Im z_0 = 0,\\ |x|^2\left(t - \frac{\Re x}{|x|^2}\right)^2 + \left(\frac{\Im x}{|x|}\right)^2, & \Im z_0 \neq 0,\end{cases}$$

   holds for each $x \in \overline{B}(z_0,\delta)$ and $t \in [0,1]$, and so

$$|1-xt| \geq \begin{cases} \{(1-z_0t)[(1-z_0t)+2t(z_0-\Re x)]\}^{1/2} \geq \delta, & \Im z_0 = 0,\\ |\Im x|/|x| \geq \delta/(\delta+|z_0|) & \Im z_0 \neq 0.\end{cases}$$

   Then use (11.38) to obtain the desired estimate.)

5. Prove the following statements:

   (a) If x is fixed in $\mathbb{C}$ and $|x| < 1$, then ${}_2F_1\left({a,\,b \atop c};x\right)$ is an analytic function of the variables a, b, and c for all finite (complex) values of a, b, and c, except for simple poles at $c \in \{0,-1,-2,\dots\}$.

   (b) ${}_2F_1\left({a,\,b \atop c};1\right)$ is an analytic function of a, b, and c for all finite values of a, b, and c such that $\Re(c-a-b) > 0$ and $c \in \mathbb{C} \setminus \{0,-1,-2,\dots\}$.

6. Give an alternative proof to the Gauss summation formula (Theorem 11.7), by using the Euler integral representation (or the technique of its proof) to firstly state the Gauss formula for $\Re c > \Re b > 0$, and then removing this constraint by analytic continuation on the parameters, using statement (b) in Exercise 5.

7. Use Gauss's summation formula and Legendre's duplication formula to show that
$$ {}_2F_1\left(\begin{matrix} -\frac{n}{2}, \; -\frac{n-1}{2} \\ \frac{2b+1}{2} \end{matrix}; 1\right) = \frac{2^n\,(b)_n}{(2b)_n}, \quad n \in \mathbb{N}, \quad \Re b > 0. $$

8. Let m and n be nonnegative integer numbers. Prove that
$$ \sum_{k=0}^{n} \binom{m}{k}\binom{m+n-k}{m}\frac{(-1)^k}{m+n+1-k} = \frac{n!}{(m+1)_{n+1}}. $$
(Hint. This sum can be written as
$$ \binom{m+n}{m}\frac{{}_3F_2\left(\begin{matrix} -n, \; -m, \; -m-n-1 \\ -m-n, \; -m-n \end{matrix}; 1\right)}{m+n+1}.) $$

9. Use the Dixon identity and the Euler reflection formula to show that
$$ \sum_{j=-\ell}^{\ell} (-1)^j \binom{2\ell}{\ell+j}\binom{2m}{m+j}\binom{2n}{n+j} = \frac{\binom{2\ell}{\ell}\binom{2m}{m}\binom{2n}{n}\binom{\ell+m+n}{m+n}}{\binom{\ell+m}{\ell}\binom{\ell+n}{\ell}}, $$
where $m, n \in \mathbb{N}_0$ and $\ell := \min\{m, n\}$.

10. Show that $y := {}_2F_1\left(\begin{matrix} a, \; b \\ c \end{matrix}; x\right)$ fulfills the hypergeometric differential equation
$$ x(1-x)\,y'' + [c-(a+b+1)x]\,y' - ab\,y = 0, \quad |x| < 1. $$

11. Prove the following hypergeometric representations of the classical orthogonal polynomials of Hermite, Laguerre, Jacobi, and Bessel (with standard normalization):
$$ H_n(x) = (2x)^n\,{}_2F_0\left(\begin{matrix} -\frac{n}{2}, \; \frac{1-n}{2} \\ - \end{matrix}; -\frac{1}{x^2}\right), $$
$$ L_n^{(\alpha)}(x) = \binom{n+\alpha}{n}\,{}_1F_1\left(\begin{matrix} -n \\ \alpha+1 \end{matrix}; x\right), $$
$$ P_n^{(\alpha,\beta)}(x) = \binom{n+\alpha}{n}\,{}_2F_1\left(\begin{matrix} -n, \; n+\alpha+\beta+1 \\ \alpha+1 \end{matrix}; \frac{1-x}{2}\right) $$
$$ = \binom{n+\alpha}{n}\left(\frac{x+1}{2}\right)^n {}_2F_1\left(\begin{matrix} -n, \; -n-\beta \\ \alpha+1 \end{matrix}; \frac{x-1}{x+1}\right), $$
$$ Y_n^{(\alpha)}(x) = {}_2F_0\left(\begin{matrix} -n, \; n+\alpha+1 \\ - \end{matrix}; -\frac{x}{2}\right). $$

These formulas hold for every $n \in \mathbb{N}_0$ and $x \in \mathbb{C}$ (with the natural definitions by continuity at the point $x = 0$ in the Hermite representation and at the point $x = -1$ in the second representation for the Jacobi polynomials).

12. Prove the following hypergeometric representations of the classical discrete orthogonal polynomials of Charlier and Meixner introduced in Exercises 3 and 4 of Chapter 4:

$$C_n^{(a)}(x) = (-a)^n \, {}_2F_0 \left(\begin{matrix} -n, \; -x \\ - \end{matrix} ; -\frac{1}{a} \right),$$

$$m_n(x;\beta,c) = (x+\beta)_n \, {}_2F_1 \left(\begin{matrix} -n, \; -x \\ -x-\beta-n+1 \end{matrix} ; \frac{1}{c} \right).$$

These formulas hold for every $n \in \mathbb{N}_0$ and $x \in \mathbb{C}$.

Appendix A

Locally Convex Spaces

This appendix presents some results on locally convex space. For a fuller presentation we recommend [49, 52].

A.1 Definitions and Basic Properties

$\mathbb{K}$ denotes either $\mathbb{R}$ or $\mathbb{C}$.

Definition A.1. *A seminorm on a vector space X over $\mathbb{K}$ is a mapping $p : X \to [0, \infty)$ obeying the following two conditions:*

(i) $p(x + y) \leq p(x) + p(y), \quad x, y \in X;$

(ii) $p(\lambda x) = |\lambda| \, p(x), \quad x \in X, \quad \lambda \in \mathbb{K}.$

A family of seminorms $\{p_\alpha\}_{\alpha \in A}$ is said to separate points if

(iii) $p_\alpha(x) = 0, \ \alpha \in A \Rightarrow \ x = 0.$

Definition A.2. *A Locally Convex Space (LCS) is a vector space X over $\mathbb{K}$ with a family $\{p_\alpha\}_{\alpha \in A}$ of seminorms separating points. The natural topology on a LCS is the weakest topology in which all the seminorms p_α are continuous and the addition is continuous. (We will refer to it as the $\{p_\alpha\}_{\alpha \in A}$–natural topology.)*

Proposition A.1. *The natural topology of a LCS is Hausdorff.*

DOI: 10.1201/9781003432067-A

A neighborhood base at 0 for the $\{p_\alpha\}_{\alpha\in A}$–natural topology in a LCS, say $W(0)$, is given by the totality of the sets of the form

$$V(0;\epsilon,\{p_{\alpha_1},\ldots,p_{\alpha_N}\}) := \{x \in W(0) \,:\, p_{\alpha_i}(x) < \epsilon,\, i = 1,\ldots,N\},$$

$\{\alpha_1,\ldots,\alpha_N\}$ being a (finite) subset of A and $\epsilon > 0$. Consequently, given a sequence $\{x_n\}_{n\geq 1}$ in $W(0)$ and $x \in W(0)$, we have

$$x_n \to x \text{ in } X \quad \Leftrightarrow \quad p_\alpha(x_n - x) \to 0, \quad \alpha \in A. \tag{A.1}$$

Definition A.3. *Two families of seminorms $\{p_\alpha\}_{\alpha\in A}$ and $\{q_\beta\}_{\beta\in B}$ in a LCS are called equivalent if they generate the same natural topology.*

Proposition A.2. *Let $\{p_\alpha\}_{\alpha\in A}$ and $\{q_\beta\}_{\beta\in B}$ be two families of seminorms in a LCS X. The following statements are equivalent:*

(i) *$\{p_\alpha\}_{\alpha\in A}$ and $\{q_\beta\}_{\beta\in B}$ are equivalent families of seminorms;*

(ii) *each p_α is continuous in the $\{q_\beta\}_{\beta\in B}$–natural topology, and each q_β is continuous in the $\{p_\alpha\}_{\alpha\in A}$–natural topology;*

(iii) *for each $\alpha \in A$, there are $\beta_1,\ldots,\beta_n \in B$ and $C > 0$ so that*

$$p_\alpha(x) \leq C \sum_{i=1}^{n} q_{\beta_i}(x), \quad x \in X;$$

and for each $\beta \in B$, there are $\alpha_1,\ldots,\alpha_n \in A$ and $K > 0$ so that

$$q_\beta(x) \leq K \sum_{i=1}^{n} p_{\alpha_i}(x), \quad x \in X.$$

A.2 Fréchet Spaces

Theorem A.1. *Let X be a LCS. The following are equivalent:*

(i) *X is metrizable (i.e., the topology in X may be defined by a metric);*

(ii) *0 has a countable neighborhood base;*

(iii) *the topology in X is generated by some countable family of seminorms.*

If $\{p_k\}_{k\geq 0}$ is a countable family of seminorms generating the topology in a LCS X, then the application $\mathrm{d} : X \times X \to [0,\infty)$ defined by

$$\mathrm{d}(x,y) := \sum_{k=0}^{\infty} \frac{1}{2^k} \frac{p_k(x-y)}{1+p_k(x-y)}, \quad x, y \in X,$$

is a metric in X and it generates the same topology in X as the family $\{p_k\}_{k\geq 0}$.

Definition A.4. *A complete metrizable LCS is called a Fréchet space.*

Recall that a complete metric space is a metric space in which every Cauchy sequence is convergent. In a metrizable LCS, whose topology is generated by the countable family of seminorms $\{p_k\}_{k\geq 0}$, a sequence $\{x_n\}_{n\geq}$ is Cauchy if and only if

$$\forall \epsilon > 0, \forall k \in \mathbb{N}_0, \exists n_0 \in \mathbb{N}_0 : \forall n, m \in \mathbb{N}_0, n, m \geq n_0 \Rightarrow p_k(x_n - x_m) < \epsilon.$$

A.3 The Inductive Limit Topology

Definition A.5. *Let X be a vector space and let $\{X_n\}_{n\geq 0}$ be a family of subspaces of X such that*

$$X_n \subseteq X_{n+1}, \quad X = \bigcup_{n=0}^{\infty} X_n.$$

Suppose that each X_n is a LCS and let $i_n : X_n \to X$ be the natural injection from X_n into X. Then the following holds:

(i) *The inductive limit topology in X of the spaces X_n is the strongest topology in X such that X is a LCS and all the maps i_n are continuous. We write*

$$X = \operatorname{ind}\lim_n X_n;$$

(ii) *if each X_{n+1} induces in X_n the given topology in X_n (i.e., X_n is a topological subspace of X_{n+1} with the relative topology), the above topology is called the strict inductive limit topology of the spaces X_n;*

(iii) *if—in addition to the conditions in* (ii)*—each X_n is a proper closed subspace of X_{n+1}, the above topology is called the hyper strict inductive limit topology of the spaces X_n.*

Theorem A.2. *Let X be a LCS endowed with the strict inductive limit topology of the LCS X_n. Then the following holds:*

(i) *The restriction of the (strict inductive limit) topology on X to each X_n is the given topology on X_n;*

(ii) *the collection of all convex sets $U \subseteq X$, $U \cap X_n$ being open in X_n for each n, is a neighborhood base at 0 in X.*

Theorem A.3. *Let X be a LCS with the strict inductive limit topology of the LCS X_n and let Y be any LCS. Then, a linear mapping $T : X \to Y$ is continuous if and only if each of the restrictions $T_n := T \,|\, X_n : X_n \to Y$ is continuous.*

Theorem A.4. *Let X be a LCS with the hyperstrict inductive limit topology of the LCS X_n. Then the following holds:*

(i) *If $\{x_n\}_{n\geq 0}$ is a sequence in X and $x \in X$, then*

$$x_n \to x \text{ in } X \quad \Leftrightarrow \quad \exists k \in \mathbb{N}_0 : x_n \in X_k \wedge x_n \to x \text{ in } X_k;$$

(ii) *if all the spaces X_n are sequentially complete, then so is X;*

(iii) *X is not a metrizable space.*

A.4 The Weak Dual Topology

Let X be a vector space over $\mathbb{K}$. The algebraic dual of X, denoted by X^*, is the set of all linear functionals $\mathbf{f} : X \to \mathbb{K}$. The action of a functional $\mathbf{f} \in X^*$ over a vector $x \in X$, $\mathbf{f}(x)$, shall be denoted by $\langle \mathbf{f}, x \rangle$. If X is endowed with a compatible topology (addition and scalar multiplication are continuous mappings), X is called a Topological Vector Space (TVS). We denote by $\mathcal{L}(X, Y)$ the set of all linear and continuous operators between two TVS X and Y. In particular, the topological dual of X is the set

$$X' := \mathcal{L}(X, \mathbb{K}) = \{\mathbf{f} \in X^* \, : \, \mathbf{f} \text{ is continuous}\}.$$

Of course, $X' \subseteq X^*$. This inclusion is actually an equality if X is a finite dimensional normed space, while it is a strict inclusion whenever X is an infinite dimensional normed space. We emphasize, however, that there are infinite dimensional TVS, X, such that the set equality $X' = X^*$ holds.

Definition A.6. *Let X be a TVS. The weak dual topology in X' is the topology in X' generated by the family of seminorms $\mathcal{S} := \{s_x \,|\, x \in X\}$, where each seminorm $s_x : X' \to [0, +\infty)$ is defined by*

$$s_x(\mathbf{f}) := |\langle \mathbf{f}, x \rangle|, \quad \mathbf{f} \in X'.$$

Endowed with the weak dual topology, X' becomes a LCS, and so a Hausdorff space. Henceforth, according with (A.1), given a sequence $\{\mathbf{f}_n\}_{n \geq 0}$ in X', we have

$$\mathbf{f}_n \to \mathbf{0} \text{ in } X' \quad \text{iff} \quad \langle \mathbf{f}_n, x \rangle \to 0, \quad x \in X.$$

Because of this property, often the name point convergence topology is given to the weak dual topology in X'. Another name which we may find in the literature is topology of convergence on the finite subsets of X. This name was given due to the fact that the collection of the sets

$$V_{X'}(\mathbf{0}; \epsilon, F) := \{\mathbf{f} \in X' \, : \, s_x(\mathbf{f}) < \epsilon, \quad x \in F\},$$

F being a finite subset of X and $\epsilon > 0$, is a neighborhood base at $\mathbf{0} \in X'$.

Definition A.7. *Let X and Y be TVS and let $T \in \mathcal{L}(X, Y)$. The dual operator (or dual mapping) of T is the linear mapping*

$$T' : Y' \to X', \quad \mathbf{g} \in Y' \mapsto T'\mathbf{g} \in X',$$

where $T'\mathbf{g} : X \to \mathbb{K}$ *is defined by*

$$\langle T'\mathbf{g}, x\rangle := \langle \mathbf{g}, Tx\rangle, \quad x \in X.$$

Theorem A.5. *Let* X *and* Y *be TVS and let* $T \in \mathcal{L}(X, Y)$. *Let* X' *and* Y' *be endowed with the weak dual topology, then* $T' \in \mathcal{L}(Y', X')$.

Bibliography

[1] W. A. Al-Salam and T. S. Chihara. Another characterization of the classical orthogonal polynomials. *SIAM J. Math. Anal.*, 3:65–70, 1972.

[2] G. E. Andrews, R. Askey, and R. Roy. *Special Functions*, volume 71 of *Encyclopedia of Mathematics and its Applications*. Cambridge University Press, Cambridge, 1999.

[3] W. Van Assche. Orthogonal polynomials, associated polynomials and functions of the second kind. *J. Comput. Appl. Math.*, 37:237–249, 1991.

[4] N. M. Atakishiyev, M. Rahman, and S. K. Suslov. On classical orthogonal polynomials. *Constr. Approx.*, 2(11):181–226, 1995.

[5] W. N. Bailey. *Generalized hypergeometric series*, volume 32 of *Cambridge Tracts in Mathematics and Mathematical Physics*. Stechert-Hafner, Inc., New York, 1964.

[6] C. Berg. Markov's theorem revisited. *J. Approx. Theory*, 78:260–275, 1994.

[7] S. Bochner. Über Sturm-Liouvillesche Polynomsysteme. *Math. Z.*, 29:730–736, 1929.

[8] J. Charris Castañeda, G. Salas, and V. Silva. Orthogonal polynomials associated with spectral problems. *Rev. Colombiana Mat.*, 25:35–79, 1992.

[9] K. Castillo. Remark on "When are all the zeros of a polynomial real and distinct?". *São Paulo J. Math. Sci.*, 16:1030–1031, 2022.

[10] K. Castillo, D. Mbouna, and J. Petronilho. On the functional equation for classical orthogonal polynomials on lattices. *J. Math. Anal. Appl.*, 515:126390, 2022.

[11] K. Castillo and J. Petronilho. Classical orthogonal polynomials revisited. *Results Math.*, 78:155, 2023.

[12] M. Chamberland. When are all the zeros of a polynomial real and distinct? *Amer. Math. Monthly*, 127:449–451, 2020.

[13] J. A. Charris, B. H. Aldana, and G. Preciado. Recurrence relations, continued fractions and determining the spectral properties of orthogonal systems of polynomials. *Rev. Acad. Colombiana Cienc. Exact. Fís. Natur.*, 27:381–421, 2003.

[14] T. S. Chihara. *An introduction to Orthogonal Polynomials.* Gordon and Breach, New York, 1978.

[15] A. Erdélyi, W. Magnus, F. Oberhettinger, and F. G. Tricomi. *Higher transcendental functions. Vol. I-III.* Robert E. Krieger Publishing Co., Inc., Melbourne, FL, 1981.

[16] K. Fan and G. Pall. Imbedding conditions for hermitian and normal matrices. *Canadian J. Math.*, 9:298–304, 1957.

[17] M. Foupouagnigni, M. Kenfack-Nangho, and S. Mboutngam. Characterization theorem of classical orthogonal polynomials on nonuniform lattices: The functional approach. *Integral Transforms Spec. Funct.*, 22:739–758, 2011.

[18] G. Freud. *Orthogonal polynomials.* Pergamon Press, Oxford-New York, 1971.

[19] J. Geronimus. Sur les polynômes orthogonaux relatifs à une suite de nombres donnée et sur le théorème de W. Hahn. (Russian. French summary). *Bull. Acad. Sci. URSS. Sér. Math. [Izvestia Akad. Nauk SSSR]*, 4:215–228, 1940.

[20] Y. L. Geronimus. On the trigonometric moment problem. *Ann. of Math.*, 47(2):742–761, 1946.

[21] W. Hahn. über die Jacobischen Polynome und zwei verwandte Polynomklassen. *Math. Z.*, 39:634–638, 1935.

[22] W. Hahn. Über höhere Ableitungen von Orthogonalpolynomen. *Math. Z.*, 43:101, 1938.

[23] M. E. H. Ismail. *Classical and quantum orthogonal polynomials in one variable*, volume 98 of *Encyclopedia of Mathematics and Its Applications.* Cambridge University Press, Cambridge, 2005.

[24] R. Koekoek, P. A. Lesky, and R. F. Swarttouw. *Hypergeometric Orthogonal Polynomials and Their q-Analogues.* Springer Monographs in Mathematics, 2010.

[25] E. Koelink. Spectral theory and special functions. In F. Marcellán R. Álvarez-Nodarse and W. Van Assche, editors, *Laredo Lectures on Orthogonal Polynomials and Special Functions*, Hauppauge, NY, 2004. Nova Science Publishers, Inc.

[26] A. N. Kolmogorov and S. V. Fomin. *Introductory real analysis.* Dover Publications, INC., New York, 1975.

[27] H. L. Krall and O. Frink. A new class of orthogonal polynomials: The Bessel polynomials. *Trans. Amer. Math. Soc.*, 65:261–264, 1949.

[28] N. N. Lebedev. *Special functions and their applications.* Prentice-Hall, Inc., Englewood Cliffs, NJ, 1965.

[29] G. Leoni. *A first course in Sobolev spaces*, volume 105 of *Grad. Stud. Math.* American Mathematical Society, Providence, RI, 2009.

[30] F. Marcellán and J. Petronilho. On the solution of some distributional differential equations: existence and characterizations of the classical moment functionals. *Integral Transform. Spec. Funct.*, 2:185–218, 1994.

[31] P. Maroni. Sur quelques espaces de distributions qui sont des formes linéaires sur l'espace vectoriel des polynômes. (French) [Some distribution spaces that are linear forms on the vector space of polynomials]. In *Orthogonal polynomials and applications (Bar-le-Duc, 1984)*, volume 1171 of *Lecture Notes in Math.*, pages 184–194, Berlin, 1985. Springer.

[32] P. Maroni. Prolégomènes à l'étude des polynômes orthogonaux semi-classiques. (French) [Prolegomena to the study of semiclassical orthogonal polynomials]. *Ann. Mat. Pura Appl.*, 4:165–184, 1987.

[33] P. Maroni. Le calcul des formes linéaires et les polynômes orthogonaux semi-classiques. (French) [Calculation of linear forms and semiclassical orthogonal polynomials]. In *Orthogonal polynomials and their applications (Segovia, 1986)*, number 1329, pages 279–290, Berlin, 1988. Springer.

[34] P. Maroni. Une théorie algébrique des polynômes orthogonaux. Applications aux polynômes orthogonaux semiclassiques (French) [An algebraic theory of orthogonal polynomials. Application to semiclassical orthogonal polynomials]. In *Orthogonal polynomials and their applications (Erice, 1990)*, IMACS Ann. Comput. Appl. Math., pages 95–130, Baltzer, Basel, 1991.

[35] P. Maroni. Variations autour des polynômes orthogonaux classiques.(french. english summary)[variations on classical orthogonal polynomials]. *C. R. Acad. Sci. Paris Sér. I Math.*, 313:209–212, 1991.

[36] P. Maroni. Variations around classical orthogonal polynomials. Connected problems. *J. Comput. Appl. Math.*, 48:133–155, 1993.

[37] P. Maroni. Fonctions eulériennes. Polynômes orthogonaux classiques. *Techniques de l'Ingénieur, traité Généralités (Sciences Fondamentales)*, A 154:1–30, 1994.

[38] P. Maroni and Z. da Rocha. A new characterization of classical forms. *Commun. Appl. Anal.*, 5:351–362, 2001.

[39] D. Mbouna. *On some Problems in the Theory of Orthogonal Polynomials.* PhD thesis, University of Coimbra, 2021.

[40] S. Mboutngam, M. Foupouagnigni, and P. Njionou Sadjang. On the modifications of semi-classical orthogonal polynomials on nonuniform lattices. *J. Math. Anal. Appl.*, 445(1):819–836, 2017.

[41] P. J. McCarthy. Characterizations of classical polynomials. *Portugal. Math.*, 20:47–52, 1961.

[42] A. F. Nikiforov, S. K. Suslov, and V. B. Uvarov. Classical orthogonal polynomials of a discrete variable on nonuniform grids. (Russian). *Dokl. Akad. Nauk SSSR*, 291(5):1056–1059, 1986.

[43] A. F. Nikiforov, S. K. Suslov, and V. B. Uvarov. *Classical orthogonal polynomials of a discrete variable. Translated from the Russian.* Springer Series in Computational Physics. Springer-Verlag, 1991.

[44] A. F. Nikiforov and V. B. Uvarov. *Special functions of mathematical physics. A unified introduction with applications. Translated from the Russian and with a preface by Ralph P. Boas. With a foreword by A. A. Samarskiĭ.* Birkhäuser Verlag, Basel, 1988.

[45] A. F. Nikiforov and V. B. Uvarov. Polynomial solutions of hypergeometric type difference equations and their classification. *Integral Transform. Spec. Funct.*, 3:223–249, 1993.

[46] E. M. Nikishin and V. N. Sorokin. *Rational approximations and orthogonality*, volume 92 of *Transl. Math. Monogr.* American Mathematical Society, Providence, RI, 1991.

[47] M. Petkovšek, H. A. Wilf, and D. Zeilberger. $A = B$. A K Peters, Ltd., Wellesley, MA, 1996.

[48] E. D. Rainville. *Special functions.* Chelsea Publishing Co., Bronx, NY, 1971.

[49] M. Reed and B. Simon. *Methods of modern mathematical physics. I. Functional analysis.* Academic Press, New York-London, 1972.

[50] R. Remmert. *Classical topics in complex function theory*, volume 172 of *Grad. Texts in Math.* Springer-Verlag, New York, 1998.

[51] R. Roy. Binomial identities and hypergeometric series. *Amer. Math. Monthly*, 94:36–46, 1987.

[52] F. Trèves. *Topological vector spaces, distributions and kernels.* Academic Press, New York-London, 1967.

[53] B. Wendroff. On orthogonal polynomials. *Proc. Amer. Math. Soc.*, 12:554–555, 1961.

[54] E. T. Whittaker and G. N. Watson. *A course of modern analysis.* Cambridge Math. Lib. Cambridge University Press, Cambridge, 1996.

Index

Printed in the United States
by Baker & Taylor Publisher Services